ACTION AND
ITS CONTROL

Technical Council on
Cold Regions Engineering Monograph

A State of the Practice Report prepared by the
Technical Council on Cold Regions Engineering
of the American Society of Civil Engineers

Edited by
Richard L. Berg and Edmund A. Wright
U.S. Army Cold Regions Research and Engineering Laboratory
Hanover, New Hampshire

AMERICAN
SOCIETY OF
CIVIL
ENGINEERS
FOUNDED
1852
®

Published by the
American Society of Civil Engineers
345 East 47th Street
New York, New York 10017-2398

The material presented in this publication has been prepared in accordance with generally recognized engineering principles and practices, and is for general information only. This information should not be used without first securing competent advice with respect to its suitability for any general or specific application.

The contents of this publication are not intended to be and should not be construed to be a standard of the American Society of Civil Engineers (ASCE) and are not intended for use as a reference in purchase specifications, contracts, regulations, statutes, or any other legal document.

No reference made in this publication to any specific method, product, process, or service constitutes or implies an endorsement, recommendation, or warranty thereof by ASCE.

ASCE makes no representation or warranty of any kind, whether express or implied, concerning the accuracy, completeness, suitability or utility of any information, apparatus, product, or process discussed in this publication, and assumes no liability therefor.

Anyone utilizing this information assumes all liability arising from such use, including but not limited to infringement of any patent or patents.

FOREWORD

The ASCE Technical Council on Cold Regions Engineering (TCCRE) established the Committee on Freezing and Thawing of Soil-Water Systems (FTSS) to "encourage development of new scientific and engineering knowledge of freezing and thawing of soil-water systems and to foster dissemination of this knowledge." To accomplish these objectives, the Committee is sponsoring the publication of a series of monographs which address major aspects of civil engineering in cold regions. Each monograph volume reviews the state-of-the-practice within a defined area through interrelated manuscripts prepared by invited authors.

This publication constitutes the third volume of the cold regions engineering monographs. Parts or all of the four papers included herein were presented during a session entitled "Frost Heave Processes and Their Control" at the ASCE National Spring Convention in Las Vegas, Nevada, in April 1982. The session was coordinated by D.R. Kane, former Chairman of the Committee, and H.P. Thomas presided during the session. Each of the papers in this monograph volume has received three positive peer reviews. All papers are eligible for discussion in the Journal of the Technical Councils of ASCE and are eligible for ASCE awards.

Richard L. Berg
Chairman, FTSS

CONTENTS

PRINCIPLES OF SOIL FREEZING AND FROST HEAVING

Duwayne M. Anderson[1], Peter J. Williams[2], Gary L. Guymon[3]
and Douglas L. Kane[4]

ABSTRACT

    This paper consists of a review of the fundamental principles of
soil freezing and thawing. Additional progress in the understanding of
frost-heaving and in the ability to perform the accurate numerical simulations of phenomena associated with soil freezing and thawing require
additional experimental and field investigations to more accurately and
more fully define and quantify these principles and their interrelationships. Frost heaving occurs inevitably when the following three conditions exist: 1) the presence of "frost susceptible soil," 2) ground
temperatures below 0°C (32°F), and 3) a source of capillary or ground
water sufficient to form and supply accumulating ice lenses in the
freezing soil. The basic principles governing the occurrence of segregated ice and concomitant frost heaving in frozen ground involve the
following: (a) the tendency of natural systems to move toward a state
of minimum energy, (b) the rate at which soil and earth materials transmit heat and moisture, (c) the liberation of the "latent heat" when
water changes from the liquid to the solid phase, (d) the redistribution
and reconfiguration of interfaces and changes in the interfacial
energies created during the resulting soil freezing or thawing, and (e)
the restructuring of the soil matrix due to the resulting soil deformation. Control or prevention of frost heaving requires the removal or
negation of one or more of the three conditions enumerated above. Once
properly understood, frost heaving can be reduced to negligible proportions or completely prevented.

INTRODUCTION

    The purpose of this paper is to provide a brief review of the
fundamental principles of soil freezing and thawing, together with an
introduction to the modern literature of ice segregation and frost
heaving. Because the phenomenology of frost heaving touches many scientific and engineering disciplines, the literature is widely dispersed.
The North American literature is accessible with little difficulty; the
European, Soviet and Asian literature is much more difficult to locate

[1]Faculty of Natural Sciences and Mathematics, State University of New
York at Buffalo, Amherst, New York 14260.
[2]Geotechnical Science Laboratories, Carleton Unversity, Ottawa,
Ontario, Canada.
[3]Department of Civil Engineering, University of California, Irvine,
California 92717.
[4]Institute of Water Resources, University of Alaska, Fairbanks, Alaska
99701.

and obtain. Fortunately, in recent years most of the important ideas and experiences have been brought together in international meetings devoted to discussions of one or more aspects of the subject. Available space does not permit a complete listing, but reference is made to many of the most useful contributions. These in turn can lead, through their references, to a more complete and detailed coverage.

It has long been observed that when a soil mass is frozen it increases in volume, often by an amount considerably greater than that which could be accounted for by the normal increase in the volume of water as it changes from the liquid to the solid state. When water crystallizes, it expands by approximately 9% of its original volume. When one takes into account the normal void ratios and the commonly encountered soil-water contents, an overall expansion of about 2 to 3% of the original soil mass would, therefore, seem to be the most reasonable expectation. For a seasonally frozen layer this would correspond to a vertical expansion on the order of two to five centimeters in typical temperate climates. Observations dating back more than one hundred years have, however, confirmed that in many soil types the vertical expansion greatly exceeds that which could be accounted for from the normal expansion of pore water on freezing. Such large vertical expansions are termed frost heaving and quite commonly range from 20 to 50 centimeters over a winter season.

A close examination of the frozen soil usually reveals that the water is frozen in the pore space and that the ice is also found in discontinuous layers, referred to as "ice lenses." The "ice lenses" appear to be more or less segregated from the soil grains. The total volume of the ice is often considerably greater than the total volume of the original pore water, and the magnitude of vertical displacement has been observed to be approximately equal to the total thickness of the segregated ice.

Frost heaving was first subjected to detailed study by Taber (86-89). He assembled experimental apparatus and conducted freezing experiments under carefully controlled conditions. He showed that, when soil is frozen from the top down under conditions permitting the upward migration of pore water from an artificial water table, continued exposure to freezing temperatures resulted in continuous vertical displacement with the appearance of a succession of segregated ice lenses as the freezing isotherm penetrated deeper and deeper into the soil material. He could observe no marked change in the rate of vertical displacement as one ice lens ceased to enlarge and a new one appeared below it. Another very important observation, frequently unmentioned and unappreciated in later discussions of frost-heaving, was that similar behavior was observed in soil materials wetted with benzene or nitrobenzene. These two liquids differ in their physical characteristics from water in that, instead of expanding, each contracts when changing from the liquid to the solid phase. This important observation provides a key to the proper understanding of ice segregation in soils and the phenomena of frost heaving. It clearly demonstrates the operation of a general principle and suggests that ice segregation and frost heaving are greatly influenced by surface and interfacial energies. Subsequent experiments and observations have confirmed and built upon this principle.

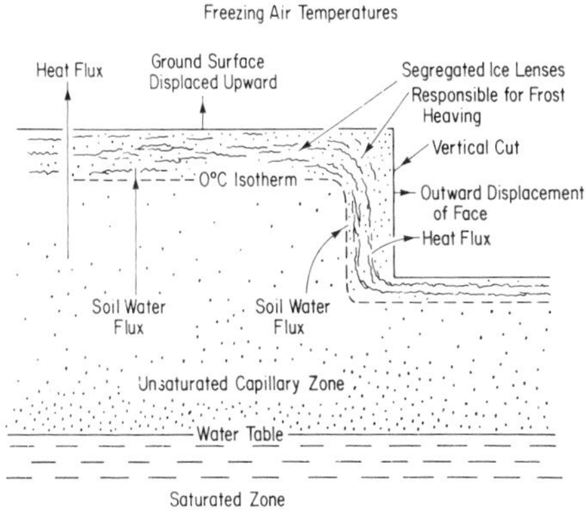

Figure 1. Schematic illustration of frost heaving.

The conditions necessary for frost heaving and a schematic diagram of the processes responsible are presented in Figure 1. Freezing air temperatures set up a thermal gradient that induces heat flow upward. As heat is extracted, freezing is initiated and growing ice crystals coalesce into planar ice lenses. Ice lens enlargement is made possible by the transport of soil water from below. The enlargement of ice lenses occurs wherever the temperature is below freezing and the rate of evolution and dissipation of the latent heat of freezing does not exceed the upward flow of soil water. Ice lens growth produces an upward displacement of the ground surface and eventually leads to measurable heaving pressures. Ice lens growth and upward displacement continue as long as a favorable relationship exists among the four principal governing factors: 1) the nature of the soil matrix, 2) the rate of heat removal, 3) the upward movement of soil water, and 4) the confining pressure. When this balanced relationship is disturbed, a new ice lens begins to form at a site where a new balance exists, usually just behind the descending 0°C isotherm.

The ground surface depicted in Figure 1 shows a vertical cut to emphasize the following important points: (a) flow lines for heat conduction exit perpendicular to the soil surface, (b) flow lines for soil water movement are, in general, parallel to the flow lines for heat conduction and (c) planar ice lenses tend to be perpendicular to the soil water flux and thermal flux flow lines. Thus it is possible for the process of frost heaving also to produce horizontal displacements. The general rule, therefore, is that displacements due to frost heaving usually are perpendicular to the ground surface. In general, the nearer the water table is to the surface, because of the larger hydraulic conductivity of wetter soils, the more readily is water transported through the partially saturated capillary zone to enlarging ice lenses behind the 0°C isotherm.

In addition to deducing these basic processes and relationships,
Tabor recognized that the ice lenses must be separated by some sort of
unfrozen boundary zone from the solid particles making up the soil fabric if the clear, segregated ice lenses he observed were to thicken and
enlarge. He reasoned that as heat was conducted upward, a portion of
this boundary layer supporting the ice lens would solidify, thickening
the lens and reducing the boundary layer thickness. The film would tend
to regain its former thickness by imbibing additional water from the
moist, unsaturated soil below. Continued extraction of moisture from
this region would, in turn, set up a hydraulic gradient that would induce the continual upward flow of water necessary to perpetuate the process. Beskow (14) enlarged upon this idea and proposed that the "capillary" behavior of soils, determined primarily by grain size distribution, determines the extent to which soils may heave when subjected to
freezing conditions. Again, the concept of interfacial tensions being a
factor in frost heaving was foreshadowed; it is implicit in the reference to capillary behavior of a porous medium composed of soil particu-
lates. These early observations and insights were subsequently confirmed, expanded and elaborated upon by others. For example, frost susceptibility was extensively investigated by Kaplar (45); the role of
surface tensions, particularly the surface tension of the ice-water
interface was examined by Gold (24), Everett (19), Miller et al. (66),
Miller (67-69), Penner (79) and Horiguchi and Miller (41), among others.

The nature, properties and behavior of the interfacial transition
zone separating the ice lenses from the soil mass envisioned by Taber
has been the subject of many subsequent investigations. Bouyoucos (15,
16) very early established that frozen soils retain a fraction of their
total water in the unfrozen state and, indirectly, he correlated the
unfrozen water with the interfacial interaction occurring between the
mineral matrix and the soil water. Similar investigations were conducted by Beskow (14) and Tsytovich (94). The calorimetric method of
determining the quantity of unfrozen water in frozen soil was refined by
Martynov (65), Nersesova and Tsytovich (72) and Williams (95-97). Additional methods of verifying the existence of the unfrozen interfacial
water films and defining the properties of the interfacial liquid were
devised by Anderson and Hoekstra (7), Anderson (4-6), and Anderson and
Tice (8).

Less direct methods which corroborate the existence of the unfrozen
interfacial water have been contributed by Hoekstra (32), Hoekstra and
Chamberlain (34), and Nakano et al. (71). As a result of these investigations, the liquid-like characteristics of the interfacial water and
frozen soil materials have been well established. Electrical conductance measurements have demonstrated the mobility of ionic species in the
interfacial zone, and the fact that water in frozen soil can be mobil-
ized by electrical or thermal gradients demonstrates the continuity and
the high molecular mobility of the interfacial water. Soil particles
that have been observed to have migrated through massive ice prove that
individual mineral particles are free of any but the most feeble and
transitory connections to the ice phase (33). Nuclear magnetic resonance data indicate that the interfacial transition zone retains its
liquid-like characteristics to very low temperatures (-40°C or lower).
The nuclear magnetic resonance data permit an estimate of the viscosity

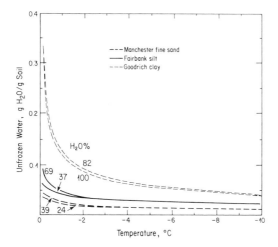

Figure 2. Phase composition of some representative soils.

of the unfrozen interfacial water. Although this method of characterizing the unfrozen film is somewhat unrealistic because of the small
physical dimensions involved, it has been concluded that the viscosity
of the unfrozen interfacial water is comparable to the viscosity of
glycerol, about 600 times greater than that of ordinary water. Measured
diffusion coefficients in frozen soils are consistent with this view.

Typical relationships between the quantity of unfrozen water in
frozen soil and temperature are shown for three representative soil
materials in Figure 2 (92). The data of Figure 2 show that the effect
of varying water and ice contents is minor. The quantity of the unfrozen interfacial water depends principally on the temperature. When
the data of Figure 2 are normalized by dividing the unfrozen water contents at any temperature by the specific surface area of the soil
materials, an important relationship emerges. All reported data giving
unfrozen water contents as a function of temperature tend to coalesce
when they are normalized by dividing them by the specific surface area
characteristic of each soil material tested. Thicknesses of the unfrozen boundary layer range from approximately two diameters of the
water molecule at temperatures of -10°C and lower to progressively
larger dimensions at temperatures approaching the melting point. Thus,
it is possible to visualize the physical arrangement of these interconnected interfacial, unfrozen water films throughout a frozen soil mass,
the thickness varying with temperature according to this characteristic
relationship.

Phase composition data such as those in Figure 2 have been obtained
for a large number of soil materials. A regression analysis of published data (8) yields, for the unfrozen water content, w_u:

$$\ln w_u = a + b \ln S + cS^d \ln \theta \qquad (1)$$

where S is the specific surface area of a given soil material in square meters per gram and θ is the temperature below $0°C$. Values of the coefficients obtained from experimental data for 11 representative soils are a = 0.2618, b = 0.5519, c = -1.449, and d = -0.264 (9). Similar regression analyses have yielded a useful relationship allowing the prediction of unfrozen water contents in frozen soils from liquid limit determinations (91).

The curves shown in Figure 2 and the general relationship obtained by normalizing them to a unit specific surface area permit many important insights to be gained in considering the properties and behavior of frozen soil. For the water in a given soil at a certain water content, consider the freezing point depression. At the freezing point of the soil water, the total water content is equivalent to the unfrozen water content w_u, since either a slight lowering of the temperature or a slight increase in the water content would create a condition in which a small quantity of ice would coexist with the unfrozen water. Looking at the same situation from a different point of view — for a frozen soil with ice and unfrozen water contents, w_i and w_u, at a given temperature, θ, below $0°C$ — removal of water from the soil by sublimation would be accomplished at the expense of w_i. The total water content w must then diminish and, as w_i approaches zero, w must equal w_u. In either of these cases, θ is equivalent to the freezing point depression at the prevailing total water content. This is a very important point for it connects purely empirical observations and measurements with thermodynamic theory. A rigorous derivation of the freezing point depression equation has been contributed by Hoekstra et al. (36, 37). Others have also discussed the matter at length (23, 50, 51, 52, 66, 90).

An accurate prediction of frost heave requires a knowledge of the basic heat and moisture transport properties of each soil in question. Considerable information on the thermal conductivity and the apparent specific heat of frozen soils has been accumulated, starting with the early work of Kerstens (48) and continuing with the work of Penner (76) and that of Slusarchuk and Watson (84). The ice content and unfrozen water content of the soil make important contributions to each of these thermal parameters. Information on the hydraulic properties of frozen soils has not been as readily obtained because of the difficulty of performing these measurements at temperatures below the freezing point of bulk water. For a few select soils, the variation in the hydraulic conductivity at temperatures below $0°C$ has been determined in the laboratory where the soil was initially saturated (17, 41). In general, as expected, the hydraulic conductivity decreases as the temperature is lowered. The decrease spans several orders of magnitude with a substantial change occurring just below the freezing point of bulk water. These observations are consistent with the data of Figure 2. As the unfrozen layer on the matrix becomes thinner, the ability of water to move within this film is reduced.

Many natural soils exist in an unsaturated condition. Very few data are available for these conditions. Although a general relationship defines the proportions of ice and unfrozen water content, given a

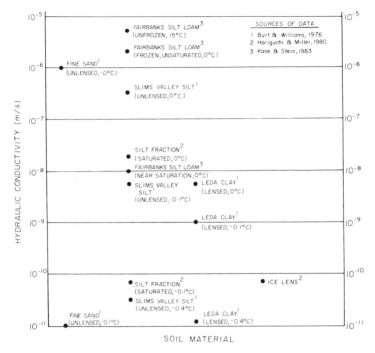

Figure 3. Typical values of hydraulic conductivity as a function of soil texture.

specified value of the total water content, it should be recognized that an array of values can exist for the hydraulic properties if an influx of water at 0°C occurs, because this unique relationship between the ice and unfrozen water content is disturbed. It should be anticipated that water would infiltrate more rapidly into a dry seasonally frozen soil than into an ice-rich seasonally frozen soil. This has been shown in results by Kane and Stein (44) performed under field conditions. When an infiltration test is performed for a sufficiently long time, the saturated hydraulic conductivity is obtained. As illustrated in Figure 3, the saturated hydraulic conductivity obtained from infiltration tests for a silt loam by Kane and Stein (44) is quite comparable to that of ice-rich soils obtained by Burt and Williams (17) and Horiguchi and Miller (39) under laboratory conditions for a silt loam at 0°C.

The preceding discussions illustrate how explanations of the phenomena of frost heaving on a theoretical basis have involved either the capillary theory on the one hand, or thermodynamic theory on the other. Actually, the two theories are not independent; the capillary theory can be derived from thermodynamical considerations. It is, therefore, not surprising that they should yield comparable predictions. This will be discussed further below. Meanwhile, it is important to emphasize, as

illustrated in Figure 1, the dynamic aspects of frost heaving which depend on the rate of heat extraction and the rate of water movement to
the freezing zone. Thermodynamic relationships and predictions that
have been developed are valid for static equilibrium conditions. Applications to dynamic, transient states are less firmly based. For this
reason the existing thermodynamic theory is not fully adequate to
describe the occurrence and processes associated with frost heaving.

To be complete, the theory of ice segregation and frost heaving
should make possible accurate and reliable predictions of rates of
heave, cumulative heave, cumulative ice-water contents, the rise and
fall of heaving pressures, and ultimate (maximum) heaving pressures.
The theory would be constructed from a set of equations that contain
expressions for the heat and soil-water fluxes induced by prevailing
thermal and water content gradients and one or more parameters to characterize the soil matrix, such as particle size, void ratio, stress distribution, interfacial curvature and dimensions, etc. Such a theory is
not yet in hand.

In the face of these difficulties, efforts have shifted to numerical modeling techniques. The basic aim of these efforts has been to de-
vise algorithms that encompass the frost-heaving processes and phenomenology and, by empirical approaches utilizing numerical techniques,
construct a comprehensive simulation model. Once a suitable model or a
set of models has been constructed and shown to adequately describe
frost heaving under fully controlled circumstances, it or they would be
expected to be utilized to predict ice segregation and frost-heaving
phenomena. It is envisioned that this would be acccomplished by extrap-
olation from fully verified circumstances, where the simulation can be
shown to be accurate, to other similar circumstances. Both finite-
element and finite-difference techniques have been investigated and
employed for this purpose (12, 23, 27, 29, 30, 38, 70, 74, 73). Considerable success has been achieved and there are grounds for optimism
that this approach will eventually lead to methods of predicting ice
segregation and frost-heaving effects reliably and with sufficient
accuracy to be useful in engineering designs.

In practical terms, the most troublesome aspects of frost-heaving
are associated with the very large increases in total soil-water content
that lead to weakening of the soil on thawing, the displacements accompanying the growth of ice lenses within the soil mass, and the buildup
of pressure whenever the normal displacements accompanying ice lens formation and growth are restrained. Some soils are more liable to exhibit
extreme behavior in this regard than others. When seasonal freeze-thaw
cycles are involved, it has been found that the most frost-susceptible
soils are those composed of particles of silt size. Gravels usually are
not susceptible to any of the undesirable characteristics of frost heaving, although ice lenses in the lower strata of alluvial channels
occasionally are observed. Clays, on the other hand, are intermediate
in their behavior. This can be explained on the basis of the processes
illustrated in Figure 1. Ice segregation and frost heaving in freezing
soil zones are dependent upon a favorable balance between the removal of
heat and the influx of soil-water, usually by capillary flow but sometimes the result of positive pore water pressures. In addition, the

soil particle sizes must be small enough to be readily rejected from
growing ice crystals in much the same manner that dissolved salts and
other solutes are excluded and rejected as liquids solidify. Crystalline solids form as individual atoms, or molecules of the liquid attach
themselves to regular positions determined by the geometry of the
crystal lattice characteristic of that material. Ice is such a material; it has hexagonal symmetry and water molecules attach one by one
during the freezing process forming a nearly perfect crystal. In this
process the ice displaces and thus rejects all foreign substances
including colloidal materials such as clays and, if freezing is slow
enough, even silts and fine sands.

Gravels are not frost susceptible because of the relatively large
mass and size of the pebbles, cobbles and rocks which make them up.
Furthermore, although gravels have a high hydraulic conductivity near
saturation, their hydraulic conductivity is very low at the low pore
water pressures associated with the freezing front. The hydraulic conductivity is much too low to accommodate significant soil-water flow
toward the freezing zone. Silts, on the other hand, have particles
small enough to be easily rejected by slowly growing ice crystals. At
the same time their hydraulic conductivity is sufficiently high that a
steady supply of water can be maintained if it is available from a nearby source.

Clays, at the other extreme, are not highly frost susceptible even
though the clay particles are very efficiently rejected by the enlarging
ice lenses (33, 42, 61, 83). The governing factor in this case is their
characteristically very low hydraulic conductivity. If the water flow
is restricted, the balance between heat removal and the influx of water
may be shifted drastically. Under these conditions, the freezing front
can rapidly invade the freezing clay. Also, because there is insufficient time for the transport of large quantities of water into the
freezing clay, the ice lenses formed usually are very thin and the displacements due to frost heaving tend to be very low. Thus, in naturally
occurring seasonal freeze-thaw, experience has shown that gravels are
the least troublesome and offer the greatest stability; silts are the
most likely to exhibit large displacements due to freeze-thaw phenomena
and are therefore classified as the most frost-susceptible soil materials; clays, on the other hand, although they usually exhibit some vertical displacements and give rise to significant heaving pressures when
confined, nevertheless are regarded as less "frost susceptible" than
silts (43, 45, 54, 81).

Frost heaving occurs inevitably whenever the following three conditions exist: 1) the presence of "frost-susceptible" soil materials; 2)
ground temperatures falling below 0°C (32°F); and 3) a source of capillary or ground water sufficient to form and supply accumulating ice
lenses in the frozen soil. Ice lenses form and enlarge steadily when
the rate of heat removal due to freezing air temperatures above the
ground surface creates a thermal gradient in the soil that exactly
matches the appropriate hydraulic gradient that, in turn, draws capillary water to the freezing front. On freezing, the water releases its
characteristic latent heat. This is a relatively large quantity compared to the heat capacities of the soil constituents and, therefore, is
the major factor in determining the balance between the two intercon-

nected fluxes: the thermal flux upward to the freezing air temperatures above and the capillary water flow from relatively warm, nearly saturated soil below, to the freezing front above. The freezing front tends to be stationary, and the ice lenses formed there tend to enlarge as long as this balance is favorable. During winter the freezing front tends to descend in incremental step-wise fashion as air temperatures steepen the thermal gradient and depletion of the soil-water below leads to a series of loci where ice lens formation is initiated. Ice lens growth proceeds for a time and then is terminated as the freezing front descends to a new, more favorable location where the process is repeated. This process is evident in the data report by Kane and Stein (44) for a seasonally frozen silt loam. When this balance is maintained, however, ice lens growth can continue for a very long time.

Control or prevention of frost heaving requires either the prevention of freezing ground temperatures, the removal and replacement of frost-susceptible soil materials, or dewatering of the ground and the prevention of water inflow. In many cases the magnitude of frost heave can be reduced significantly by increasing the overburden pressure. Once these principles are understood, then frost heaving can be reduced to negligible proportions or completely prevented in any specific instance once it is determined how one or all of these factors can be controlled. This should be properly understood and taken into account in the early design stages of all ground-supported structures or structures involving earthworks, for prevention of damage associated with frost-heaving is far easier and more economical than post-construction correction and repair.

In recent years engineering structures, such as underground containers for liquefied natural gas, buried pipelines, and the introduction of engineering techniques for temporary, artificial freezing of soil materials to achieve stabilization, have brought new dimensions into the consideration of the phenomenology of soil freezing. In these instances the major complication is that the freezing period is very much longer than the natural freezing period associated with seasonal change. The principal consequence of this is to extend the group of "frost-susceptible" materials to include both coarser and finer grained materials, even dirty gravels and clays.

The rate of penetration of a freezing front falls off roughly as the square root of time. This makes it virtually certain that, after a time, the necessary favorable balance between heat and water flux will be achieved so that, at some point, ice lens formation and growth can occur. Whenever displacement by consolidation or upheaval is restricted or constrained, potentially large internal pressures can develop. The extent to which very large pressures associated with ice segregation may be built up has been investigated both theoretically and experimentally under controlled laboratory conditions (18, 19, 20, 35-37, 49, 56, 69, 75-77, 82, 85). Theoretical predictions of maximum ice segregation pressures that may be built up during frost heaving when displacement is prevented are based securely on the fundamental principle that water, like all substances, spontaneously seeks a state of minimum energy. In general, the state of lowest energy corresponds to the lowest temperature accessible. The imposition of increasing pressure tends to raise

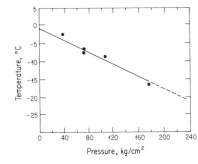

Figure 4. Relationship between temperature and maximum observed heaving pressure for montmorillonite clay.

the energy of water in the solid state and by "pressure melting" to establish a new point of equilibrium (23, 26, 36, 37). Thus, a quasi-equilibrium condition can be envisioned that involves a dynamic equilibrium between the unfrozen, interfacial water and the ice phase. This equilibrium is governed by a series of factors. The most pronounced of these is the prevailing temperature. Equally important is the pressure. Other important factors are the quantity of dissolved substances present in the interfacial water (11, 103) and the configuration of the ice/water interface. In principle, the interrelationships can be expressed by means of a special phase diagram (5, 6, 36, 37). When temperature and pressure are considered alone, a form of the Clausius-Clapeyron equation may be derived. If this is done rigorously, partial molar volumes and partial molar enthalpies are involved. These quantities are generally not available and are extremely difficult to determine accurately. Consequently, the Clausius-Clapeyron equation has been used in its conventional form to obtain approximate predictions and to test the correlation with experimental determinations (82). The result is shown in Figure 4 where it can be seen that the agreement is remarkably good. Thus it appears that, when long-term freezing is involved, clays do indeed become frost susceptible and very large ice segregation pressures can be anticipated under certain conditions. Although these high pressures may be relieved by relatively small displacements, this is an extremely important and often overlooked aspect of frost heaving by those whose background and experience have involved only seasonal freeze-thaw phenomena.

Given this understanding, the data shown in Figure 3, and the concept of extensive, interconnected, unfrozen, interfacial water films throughout a frozen soil mass whose thickness varies with temperature in the manner illustrated in Figure 2, it becomes clear that, although initial ice formation occurs just behind the 0°C isotherm, freezing continues throughout the soil mass as the temperature is lowered and water continues to be supplied by the network of unfrozen water films (5, 6, 9, 26, 56, 68, 69). We refer to this as "secondary heaving"; others have preferred to refer to this occurrence simply as water redistribution in partially frozen ground (for example, 55). Miller, who originated the term, uses it somewhat differently. The important point is the concept of water being in a condition of continual accommodation to changing temperature and stress fields. This gives rise to the phenomenon of plastic deformation and creep of frozen ground (see for example, 3).

Because of the free mobility of water within the network of unfrozen water interfacial layers, ice lenses behind the freezing front
may continue to enlarge slowly as the temperature falls. Recent evidence of this has been presented by Mackay (64). As the soil temperature falls below about $-5°C$, the unfrozen water interfaces become very
thin. This severely restricts the quantities of water that can be
transported. In theory, ice lens growth can continue until about
$-50°C$. At this point, the mobility of the individual water molecules
for all practical purposes becomes insignificant.

At all points where individual water molecules attach themselves to
the ice lattice, an increment of latent heat is released. The magnitude
of this increment depends upon the difference between the energy state
of the pure crystalline ice on the one hand, and the unfrozen, interfacial liquid on the other. The energy involved in the "work" of frost
heaving is derived from this difference. Soil consolidation and compaction to very high densities can be achieved in localized areas and
ground displacements can occur to varying degrees depending on the
physical constraints imposed. it has been widely observed that the
imposition of loads greatly reduces the magnitude of vertical displacements due to frost heaving (1, 2, 31, 38, 75, 81, 98, 101, 102). The
causes probably involve a reduction of the hydraulic conductivity
brought about by compaction of the soil at and near the freezing zone
and the pressure dependency of the relationship between unfrozen water
content and temperature. As the pressure increases (because of pressure
melting), at a given temperature, the unfrozen water content increases.
This in turn causes an increase in the pore water pressure (or decrease
in the soil water tension). Thus the hydraulic gradient is reduced and
water flows to the freezing zone more slowly. Surcharge loading cannot
completely stop the process of frost heaving (79), until the pressures
indicated in Figure 3 (82) are reached or exceeded.

The phenomenon of frost-heaving is geographically widespread. More
than half the land area of the earth may be involved. The seasonal
freeze-thaw phenomenon is an annual occurrence in the terrestrial
temperate zones. It also occurs at high elevations through the world.
In the polar regions, the ground is permanently frozen at depth, but in
many places there is an "active layer" at the surface that experiences
annual freeze-thaw regimes (62). Within the active layer, frost heaving
is common and may occur both at the upper and lower surfaces of the
active layer (59, 60). Frost heaving under these conditions has been
carefully measured and studied by Mackay et al. (63) and Mackay (64).
Although water redistribution within permafrost and the active layer
must be viewed as continually in progress, it is generally observed that
the ice content of the frozen ground diminishes with increasing depth.
Occasional, massive bodies of ground ice may result from positive pore
water pressures and be formed by injection. Others may be buried re-
licts. Still others may result from the formation of ice wedges and
pingos (13, 62).

Mention was made earlier of the trend toward numerical simulation
of the processes and phenomena associated with ice segregation and frost
heaving. The interrelationships between the simultaneous flux of thermal energy and water have been extensively studied (21, 27, 39, 40, 46,
52, 53, 57, 58, 70, 73, 74, 100). This coupled process of heat extrac-

tion and moisture supply has been identified throughout this article as the fundamental process determining the rate and extent of frost-heaving, given the three basic requirements: below-freezing temperatures, frost-susceptible soil materials, and an adequate supply of soil moisture. Considerable progress has been made in modeling coupled heat and moisture flow. This in turn, has led to elaborate numerical simulations of the overall frost-heaving phenomenon.

Although no one simulation technique has emerged as completely satisfactory, vigorous activity is rapidly improving the several approaches currently being exploited. The magnitude of the difficulties obstructing further progress should not, however, be underestimated; several problems in the routine application of numerical models remain the subject of debate. Although most authors of current models use similar equations of state, i.e. fluid continuity and Darcy's law to describe water transport together with the classical heat equation to describe heat transport, there are considerable differences in the construction of the various numerical models, primarily depending on how latent heat effects and nonlinear model parameters are dealt with. Some have attempted to use rather elaborate concepts describing the ice segregation process itself and the effects of overburden pressure while others have attempted to use simpler concepts and devices. We are not agreed on an optimal number of parameters to incorporate in a "best" model or algorithms to introduce accepted knowledge and understanding into a fully adequate numerical model. Furthermore, it is not certain that a purely deterministic model will ever be adequate. It is well known that soil parameters are subject to large measurement errors; unsaturated hydraulic conductivity for instance may have a coefficient of variation of around 500%. Variation in the critical parameters that govern ice segregation, such as freezing boundary conditions, hydraulic conductivity, and thermal conductivity, may have quite a large effect on computed frost heave (29, 30). In pursuing the objective of this approach, additional effort to verify mathematical models with laboratory and, particularly, field data is required. For only through the exhaustive verification of a number of models can confidence in their application to engineering analysis and design be established. Notwithstanding present difficulties, it is to be expected that eventually one or more numerical simulation schemes will be brought to a high degree of refinement and utility and that these schemes will be of steadily increasing value.

APPENDIX 1 - REFERENCES

1. Aitken, G.W., "Reduction of Frost Heave by Surcharge Stress," U.S. Army
    Cold Regions Research and Engineering Laboratory (CRREL) Technical
    Report 184, 1976.

2. Aitken, G.W. and Berg, R.L., "Some Passive Methods of Controlling Geo-
    cryological Conditions in Roadway Construction," Permafrost: North
    American Contribution to the Second International Conference, Yakutsk,
    U.S.S.R., National Academy of Sciences, Washington, 1973, pp. 581-586.

3. Andersland, O.B. and Anderson, D.M., Eds., Geotechnical Engineering for
    Cold Regions, McGraw-Hill, New York.

4. Anderson, D.M., "Phase Composition of Frozen Montmorillonite-Water
    Mixtures From Heat Capacity Measurements," Soil Science Society of
    America, Proceedings, Vol. 30, 1966, pp. 670-675.

5. Anderson, D.M., "The Interface Between Ice and Silicate Surfaces,"
    Journal of Colloid and Interface Science, Vol. 25, 1967, pp. 174-191.

6. Anderson, D.M., "Ice Nucleation and the Substrate-Ice Interface,"
    Nature, Vol. 216, 1967, pp. 563-566.

7. Anderson, D.M. and Hoekstra, P., "Migration of Interlamellar Water
    During Freezing and Thawing of Wyoming Bentonite," Soil Science Society
    of America, Proceedings, Vol. 29, 1965, pp. 498-504.

8. Anderson, D.M. and Tice, A.R., "Predicting Unfrozen Water Contents in
    Frozen Soils from Surface Area Measurements," Highway Research Record,
    No. 393, 1972, pp. 12-18.

9. Anderson, D.M., Tice, A.R. and Banin, A., "Prediction of Unfrozen Water
    Contents in Frozen Soils from Liquid Limit Determinations," Symposium
    on Frost Action on Roads, Proceedings Summary and Discussion, Oslo,
    Norway, October 1-3, 1973.

10. Anderson, D.M. and Morgenstern, N.R., "Physics, Chemistry and Mechanics
    of Frozen Ground," Permafrost: North American Contribution to the
    Second International Conference, Yakutsk, U.S.S.R., National Academy of
    Sciences, Washington, 1973, pp. 257-288.

11. Banin, A. and Anderson D.M., "Effects of Salt Concentration Changes
    During Freezing on the Unfrozen Water Content of Porous Materials,"
    Water Resources Research, Vol. 10, No. 1, 1974.

12. Berg, R.L., Guymon, G. and Gartner, K.E., "A Mathematical Model to
    Predict Frost Heave," Proceedings, International Symposium on Frost
    Action in Soils, Lulea, Sweden, February 1977, University of Lulea,
    Vol. 2, 1977, pp. 92-109.

13. Berg, R.L. and Smith, N., "Encoutering Massive Ground Ice During Road Construction in Central Alaska," Permafrost: North American Contribution to the Second International Conference, Yakutsk, U.S.S.R., National Academy of Sciences, Washington, 1973.

14. Beskow, G., "Soil Freezing and Frost Heaving with Special Attention to Roads and Railroads," Swedish Geological Society, Series C, 375, 26th Yearbook, 1935.

15. Bouyoucos, G.J. and McCool, M.M., "Further Studies on the Freezing Point Lowering of Soils," Michigan Agricultural Experiment Station Technical Bulletin No. 31, 1916, 51 p.

16. Bouyoucos, G.J., "Classification and Measurement of the Different Forms of Water in the Soil by Means of the Dilatometer Method," Michigan Agricultural Experiment Station Technical Bulletin No. 36, 1917, 48 p.

17. Burt, T.P. and Williams, P.J., "Hydraulic Conductivity in Frozen Soils," Earth Surface Processes, Vol. 1, John Wiley & Sons, Ltd, 1970, pp. 349-360.

18. Chamberlain, E., "The Mechanical Behavior of Frozen Earth Materials Under High Pressure: Triaxial Test Conditions," Geotechnique, Vol. 22, No. 3, 1972, pp. 469-483.

19. Everett, D.H., "Thermodynamics of Frost Damage to Porous Solids," Transactions, Faraday Society, Vol. 57, 1961, pp. 1541-1551.

20. Freden, S. and Stenberg, L., "Frost Heave Tests on Tills with an Apparatus for Constant Heat Flow," Proceedings, 2nd International Symposium on Ground Freezing, The Norwegian Institute of Technology, Trondheim, Norway, 1980, pp. 760-771.

21. Fukuda, M., Orhun, A. and Luthin, J.N., "Experimental Studies of Coupled Heat and Moisture Transfer in Soils During Freezing," Cold Regions Science and Technology, Vol. 3, No. 2 and 3, 1980, pp. 223-232.

22. Gilpin, R.R., "A Model of the 'Liquid-Like' Layer Between Ice and a Substrate with Application to Wire Regelation and Particle Migration," Journal of Colloid and Interface Science, Vol. 68, No. 2, 1979, pp. 235-251.

23. Gilpin, R.R., "A Model for Prediction of Ice Lensing and Frost Heave in Soils," Submitted to Water Resources Research, 1980.

24. Gold, L.W.," A Possible Force Mechanism Associated with the Freezing of Water in Porous Materials," Highway Research Board Bulletin, Vol. 168, 1957, pp. 65-72.

25. Groenevelt, P.H. and Kay, B.D., "Water and Ice Potentials in Frozen Soils," Water Resources Research, Vol. 13, No. 2, 1977, pp. 445-449.

26. Groenevelt, P.H. and Kay, B.D., "Pressure Distribution and Effective Stress in Frozen Soils," Second International Symposium on Ground Freezing, Trondheim, June 24-26, Norwegian Institute of Technology, 1980, pp. 597-610.

27. Guymon, G. and Luthin, J.N., "A Coupled Heat and Moisture Transport Model for Arctic Soils," Water Resources Research, Vol. 10, No. 5, 1974.

28. Guymon, G., Hromadka, T.V. and Berg, L., "A One-Dimensional Frost Heave Model Based Upon Simulation of Simultaneous Heat and Water Flux," Cold Regions Science and Technology, Vol. 3, No. 2 and 3, 1980.

29. Guymon, G., Berg, R.L., Johnson, T.C. and Hromadka, T.V., "Results from a Mathematical Model of Frost Heave," Transportation Research Record 809, 1981, pp. 2-6.

30. Guymon, G., Harr, M.E., Berg, R.L. and Hromadka, T.V., "A Probabilistic-Deterministic Analysis of One-Dimensional Ice Segregation in a Freezing Soil Column," Cold Regions Science and Technology, Vol. 5, 1981, pp. 127-140.

31. Hill, D.W. and Morgenstern, N.R., "Influence of Load and Heat Extraction on Moisture Transfer in Freezing Soils," Proceedings, International Symposium on Frost Action in Soils, Lulea, Sweden, University of Lulea, Vol. I, 1977, pp. 76-91.

32. Hoekstra, P., "Conductance of Frozen Bentonite Suspensions, "Soil Science Society of America Proceedings, Vol. 29, 1965, pp. 519-522

33. Hoekstra, P., "The Physics and Chemistry of Frozen Soils," Highway Research Board, Special Report 103, 1969, pp. 78-90.

34. Hoekstra, P. and Chamberlain, E., "Electro-osmosis in Frozen Soil," Nature, Vol. 203, 1968, pp. 1406.

35. Hoekstra, P., Chamberlain, E. and Frate, A., "Frost Heaving Pressures," Highway Research Board Record, Vol. 101, 1965, pp. 28-38.

36. Hoekstra, P., Low, P.F. and Anderson, D.M., "Some Thermodynamic Relationships for Soil at or Below Freezing Point: 1. Freezing Point Depressions and Heat Capacity," Water Resources Research, Vol. 4, No. 2, 1968, pp. 379-394.

37. Hoekstra, P., Low, P.F. and Anderson, D.M., "Some Thermodynamic Relationships for Soil at or Below the Freezing Point: 2. Effects of Temperature and Pressure on Unfrozen Soil Water," Water Resources Research, Vol. 4, No. 3, 1968, pp. 541-544.

38. Hopke, S., "A Model for Frost Heave Including Overburden," Cold Regions Science and Technology, Vol. 3, No. 2 and 3, 1980, pp. 111-127.

39. Horiguchi, K., "Relations Between the Heave Amount and the Specific Surface Area of Powdered Materials," Low Temperature Science, Ser. A, Vol. 33, 1975, pp. 237-242.

40. Horiguchi, K., "Effects of the Rate of Heat Removal on the Rate of Frost Heaving," Engineering Geology, Vol. 13, 1979, pp. 63-71.

41. Horiguchi, K. and Miller, R.D., "Experimental Studies with Frozen Soil in an Ice Sandwich Permeameter," Cold Regions Science and Technology, Vol. 3, 1980, pp. 177-183.

42. Jackson, K.A., Chalmers, B. and Uhlamm, D.R., "Particle Sorting and Stone Migration Due to Frost Heave," Science, Vol. 152, 1966, pp. 545-546.

43. Jones, R.H., "Development and Application of Frost Susceptibility Testing," Proceedings, 2nd International Symposium on Ground Freezing, Norwegian Institute of Technology, June 24-26, 1980.

44. Kane, D.L. and Stein, J., "Water Movement into Seasonally Frozen Soils," Paper submitted to Water Resources Research, 1982.

45. Kaplar, C.W., "New Experiments to Simplify Frost Susceptibility Testing of Soils," Highway Research Record, Vol. 215, 1968, pp.48-59.

46. Kay, B.D. and Groenvelt, P.H., "On the Interaction of Water and Heat Transfer in Frozen Soils: 1. Basic Theory the Vapor Phase," Soil Science Society of America, Proceedings, Vol. 38, 1974, pp. 395-400.

47. Kay, B.D., Sheppard, M.I. and Loch, J.P.G., "A Preliminary Comparison of Simulated and Observed Water Redistribution in Soils Freezing Under Laboratory and Field Conditions," Proceedings, International Symposium on Frost Action in Soils, Lulea, Sweden, University of Lulea, 1977, pp. 29-41.

48. Kerstens, M., "Thermal Properties of Soils," University of Minnesota Engineering Experiment Station, Bulletin No. 28, 1949.

49. Kinosita, S., "Heave Force of Frozen Soil," Low Temperature Science, Kitami Tech. College Ser. A-21, 1962.

50. Kinosita, S. and Takeshi, I., "Freezing Point Depression in Moist Soil," Proceedings, 2nd International Symposium on Ground Freezing, Norwegian Institute of Technology, 1980, pp. 640-646.

51. Koopmans, R.W.R. and Miller, R.D., "Soil Freezing and Soil Water Characteristic Curves," Soil Science Society of America, Proceedings, Vol. 30, 1966, pp. 680-685.

52. Loch, J., "Thermodynamic Equilibrium Between Ice and Water in Porous Media," Soil Science, Vol. 126, 1978, pp. 77-80.

53. Loch, J., "Influence of the Heat Extraction Rate on the Ice Segregation Rate of Soils," Frost i Jord, Vol. 20, 1979, pp. 19-30.

54. Loch, J., "Suggestions for an Improved Standard Laboratory Test for Frost Heave Susceptibility of Soils," Frost i Jord, Vol. 20, 1979, pp. 33-38.

55. Loch, J. and Kay, B.D., "Water Redistribution in Partially Frozen, Saturated Silt Under Several Temperature Gradients and Over-burden Loads," Soil Science Society of America, Proceedings, Vol. 42, 1978, pp. 400-406.

56. Loch, J. and Miller, R.D., " Test of the Concepts of Secondary Frost Heaving," Soil Science Socity of America, Proceedings, Vol. 39, No. 6, 1975, pp. 1036-1041.

57. Luthin, J. and Taylor, G., "Numerical Results of Coupled Heat-Mass Flow During Freezing and Thawing," Proceedings, 2nd International Conference on Soil Water problems in Cold Regions, Edmonton, Alberta, Canada, Sept. 1-2, 1976, pp. 155-172.

58. Luthin, J. and Taylor, G., "A Model for Coupled Heat and Moisture Transfer During Soil Freezing," Canadian Geotechnical Journal, Vol. 15, 1978, pp. 548-555.

59. Mackay, J.R., "Freezing Processes at the Bottom of Permafrost: Tuktoyaktuk Peninsula, District of Mackenzie," Geological Survey of Canada, Paper 75-1A, 1975, pp. 471-474.

60. Mackay, J.R., "Ice Segregation at Depth in Permafrost," Geological Survey of Canada, Paper 76-1A, 1976, pp. 287-288.

61. Mackay, J.R., "Uplift of Objects by an Upfreezing Ice Surface," Canadian Geotechnical Journal, Vol. 15, No. 3, 1978, pp. 609-613.

62. Mackay, J.R. and Black, R.F., "Origin, Composition and Structure of Perennially Frozen Ground and Ground Ice: A Review," Permafrost: North American Contribution to the 2nd International Conference, Yakutsk, U.S.S.R., National Academy of Sciences, Washington, 1973, pp. 185-192.

63. Mackay, J.R., Ostrick, J., Lewis, J. and Mackay, D.K., "Frost Heave at Ground Temperatures Below 0 degrees C, Inuvik, Northwest Territories," Geological Survey of Canada, Paper 79-1A, 1979, pp. 403-405.

64. Mackay, J.R., "Active Layer Slope Movement in a Continuous Permafrost Environment, Garry Island, Northwest Territories, Canada," Canadian Journal of Earth Sciences, Vol. 18, 1981, pp. 1666-1680.

65. Martynov, G.S., "Calorimetric Method of Determining the Quantity of Unfrozen Water in Frozen Soil," Data on the Principles of the Study of Frozen Zones on the Earth's Crust, L.A. Meister, ed., Issue III. Academy of Sciences, U.S.S.R., V.A. Obruchev, Institute of Permafrost Studies, Moscow, National Research Council of Canada, Ottawa, Technical Translation 1088, 1956.

66. Miller, R.D., Baker, J.H. and Kolaian, J.H., "Particle Size, Overburden Pressure, Pore Water Pressure and Freezing Temperature of Ice Lenses in Soil," Transactions, International Congress Soil, Science, Vol. 1, 1960, pp. 122-129.

67. Miller, R.D., "Soil Freezing in Relation to Pore Water Pressure and Temperature," Permafrost: North American Contribution to the 2nd International Conference on Permafrost, Yakutsk, U.S.S.R., 1973, pp. 344-352.

68. Miller, R.D., "Lens Initiation in Secondary Heaving," Proceedings, International Symposium on Frost Action in Soils, Lulea, Sweden, 1977.

69. Miller, R.D., "Frost Heaving in Non-Colloidal Soils," 3rd International Conference of Permafrost, Edmonton, Alberta, Canada, 1978, pp. 707-713.

70. Miller, R.D. and Koslow, E.E., "Computation of Rate of Heave Versus Load Under Quasi-Steady State," Cold Regions Science and Technology, Vol. 3, No. 2 and 3, 1980, pp. 243-252.

71. Nakano, Y., Martin, R.J. and Smith, M., "Ultrasonic Velocities of the Dilatational and Shear Waves in Frozen Soils," Water Resources Research, Vol. 8, 1972, pp. 1024-1030.

72. Nersesova, Z.A. and Tsytovich, N.A., "Unfrozen Water in Frozen Soils," Permafrost: Proceedings of an International Conference. National Academy of Sciences, Washington, 1966, pp. 230-234.

73. Outcalt, S.I., "A Simple Energy Balance Model of Ice Segregation," Cold Regions Science and Technology, Vol. 3 No. 2 and 3, 1980, pp. 145-152.

74. O'Neill, K. and Miller, R.D., "Numerical Solutions for Rigid Ice Model of Secondary Frost Heave," 2nd International Symposium on Ground Freezing. Norwegian Institute of Technology, June 24-26, 1980, pp. 656-669.

75. Penner, E., "Heaving Pressures in Soils During Unidirectional Freezing," Canadian Geotechnical Journal, Vol. IV, No. 4, 1967, pp. 398-408.

76. Penner, E., "Frost Heaving Forces in Leda Clay," Canadian Geotechnical Journal, Vol. 7, No. 8, 1970, pp. 8-16.

77. Penner, E., "Thermal Conductivity of Frozen Soils," Canadian Journal of Earth Sciences, Vol. 7, No. 3, 1970, pp. 982-987.

78. Penner, E., "Fundamental Aspects of Frost Action," 2nd International Symposium on Frost Action in Soils, Lulea, Sweden, University of Lulea, Vol. 2, 1977, pp. 17-28.

79. Penner, E. and Ueda, T., "The Dependence of Frost Heaving on Load Application," International Symposium on Frost Action in Soils, Lulea, Sweden, University of Lulea, Vol. 1, 1977, pp. 92-101.

80. Penner, E. and Ueda, T., "A Frost-Susceptibility Test and a Basis for Interpreting Heaving Rates," Proceedings, 3rd International Conference on Permafrost, Edmonton, Alberta, Canada, National Research Council of Canada, Ottawa, 1978, pp. 721-727.

81. Penner, E. and Ueda, T., "Effects of Temperature and Pressure on Frost Heaving," Engineering Geology, Vol. 13, No. 1-4, 1979, pp. 29-39.

82. Radd, F.J. and Oertle, D.H., "Experimental Pressure Studies of Frost Heave Mechanisms and the Growth-Fusion Behavior of Ice in Soils and Glaciers," Permafrost: North American Contribution to the 2nd International Conference, Yakutsk, U.S.S.R., 1973, pp. 377-384.

83. Romkens, M.J.M. and Miller, R.D., "Migration of Mineral Particles in Ice with a Temperature Gradient," Journal of Colloid and Interface Science, Vol. 42, 1973, pp. 103-111.

84 Slusarchuk, W.A. and Watson, G.H., "Thermal Conductivity of Some Ice Rich Permafrost Soils," Canadian Geotechnical Journal, Vol. 12, 1975, pp. 413-424.

85. Stenberg, L., "Frost Heave Studies by Natural Freezing," 2nd International Symposium on Ground Freezing, Norwegian Institute of Technology, June 24-26, 1980, pp. 784-794.

86. Taber, S., "The Growth of Crystals Under External Pressure," American Journal of Science, 4th Ser., Vol XLI, 1916, pp. 532-556.

87. Taber, S., "Frost Heaving," Journal of Geology, Vol. 37, No. 1, 1929, pp. 428-461.

88. Taber, S., "Freezing and Thawing of Soils as Factors in the Destruction of Road Pavements," Public Roads, Vol. VII, No. 6, 1930, pp. 113-132.

89. Taber, S., "The Mechanics of Frost Heaving," Journal of Geology, Vol. 38, 1930, pp. 303-317.

90. Takagi, S., "Theory of Freezing-Point Depression with Special Reference to Soil Water," Permafrost: Proceedings of an International Conference, Lafayette, Indiana, 1963, pp. 216-224.

91. Tice, A.R., Anderson, D.M. and Banin, A., "The Prediction of Unfrozen Water Contents in Frozen Soils from Liquid Limit Determinations," U.S. Army Cold Regions Research & Engineering Laboratory, CRREL Report 76-8, 1976.

92. Tice, A.R., Burrous, C.M. and Anderson, D.M., "Determination of Unfrozen Water in Frozen Soil by Pulsed Nuclear Magnetic Resonance," Proceedings, 3rd International Conference on Permafrost, Edmonton, Alberta, Canada, National Research Council of Canada, Ottawa, 1978, pp. 149-155.

93. Tice, A.R., Burrous, C.M. and Anderson, D.M., "Phase Composition Measurements on Soils at Very High Water Contents by the Pulsed Nuclear Magnetic Resonance Technique," Transportation Research Record, Vol. 675, 1978, pp. 11-14.

94. Tsytovich, N.A., "On the Theory of the Equilibrium State of Water in Frozen Soils," Izv. AN SSSR Sev. Geogr., Vol. 9, 1945, pp. 5-6.

95. Williams, P.J., "Specific Heats and Unfrozen Water Content of Frozen Soils," National Research Council, Canada, Associate Committee on Soil and Snow Mechanical Techniques, Mem. No. 76, 1963, pp. 109-126.

96. Williams, P.J., "Experimental Determination of Apparent Specific Heats of Frozen Soils," Geotechnique, Vol. XIV, No. 2, 1964, pp. 133-142.

97. Williams, P.J., "Unfrozen Water Content of Frozen Soils and Soil Moisture Suction," Geotechnique, Vol. 14, No. 3, 1964, pp. 231-246.

98. Williams, P.J., "Pore Pressures at a Penetrating Frost Line and Their Prediction," Geotechnique, Vol. 16, No. 3, 1966, pp. 187-208.

99. Williams, P.J., "The Nature of Freezing Soils and its Field Behavior," Norwegian Geotechnical Institute, Publication 72, 1967.

100. Williams, P.J., "Thermodynamic Conditions for Ice Accumulation in Freezing Soils," International Symposium on Frost Action in Soils, Lulea, Sweden, University of Lulea, Vol. 1, 1977, pp. 42-53.

101. Williams, P.J., Pipelines and Permafrost, Longman, London, 1979, p. 103.

102. Yong, R.N., "Heave and Heaving Pressures in Frozen Soils," Canadian Geotechnical Journal, Vol. 8, 1971, pp. 272-282.

103. Yong, R.N. and Cheung, C.H., "Prediction of Salt Influence on Unfrozen Water Content in Frozen Soils," Journal of Engineering Geology, Vol 13, 1979, pp. 137-155.

DESIGNING FOR FROST HEAVE CONDITIONS

by

Frederick E. Crory[1], M.ASCE, Ralph M. Isaacs[2], Edward Penner[3],
Frederick J. Sanger[4], F.ASCE (Life Member), and James F. Shook[5],
F.ASCE

ABSTRACT

Current approaches and information concerning frost heave design
for foundations, buried pipelines, artificial ground freezing,
in-ground cryogenic liquid storages, and pavements and other roadbeds
are discussed, and key research needs are summarized. It is concluded
that U.S. building codes should be revised to take effects of tangential frost heave forces into account. Improvements are also needed
in frost susceptibility criteria and tests and in heat transfer
analysis techniques. A frost heave design approach has been developed
for buried pipelines operated at below-freezing temperatures that is
applicable for silty soils, but it is suspected that heave behavior in
clays may be very different. Most important areas of needed research
on artificial ground freezing and in-ground storage of cryogenic
liquids are the rate of growth, shape, and spread of the frost bulb,
surface and lateral heave effects, and fissuring. In designing
pavements and other road beds for frost heave, design techniques
should be applied on a selective basis for the most economical
results.

INTRODUCTION (Penner)

Ground freezing introduces special geotechnical considerations in
the design of engineering structures. Freezing may originate either
from naturally occurring seasonal frost or it may be due to artificial
freezing imposed by the operation of the structure itself. The damage
to the structures that must be avoided usually arises from displacements - frost heave - initiated by the ice/water phase change in the
soil.

---

[1]Research Civil Engineer, Geotechnical Research Branch, Experimental
Engineering Division, U.S. Army Cold Regions Research and Engineering
Laboratory, Hanover, NH.

[2]Manager, Frost Heave Design, Northwest Alaskan Pipeline Company,
Irvine, California.

[3]Head, Geotechnical Section, Division of Building Research, National
Research Council of Canada, Ottawa, Ontario, Canada.

[4]Consulting Engineer, Orange City, Florida.

[5]Principal Engineer, Asphalt Institute, College Park, Maryland.

In subsequent sections of this paper we discuss (a) design considerations for various types of foundations, pavements and other road beds subjected to seasonal frost, (b) frost heave due to artificial freezing for construction purposes and to in-ground storage of cryogenic liquids, and (c) the nature of the problem and the design approach to a buried chilled pipeline.

It is apparent that a better understanding of frost heaving processes developed in recent years has advanced the state of the art in all areas of structural design.

Phenomena Associated with Frost Heaving

Heaving can be attributed mainly to the growth of ice lenses. In the early days, however, the popular belief was that frost heaving resulted only from the expansion associated with the ice/water phase change. Ice lenses usually form parallel to the isothermal freezing plane, although vertical cracks present in the soil will also fill with ice. Heave forces and heave displacements are always in the direction of heat flow. Under natural conditions this is normal to the horizontal ground surface. Work must be done to lift the overburden and any surface surcharges. Normally water is obtained from the unfrozen soil or the water table and so further work must be done to induce and maintain the suction gradient. The energy required for these two processes is made available at the ice/water interface of the actively growing ice lenses.

At least four major contributions in the last 50 years have led to a more complete understanding of frost action processes as applied to the design of structures for engineering purposes. These were 1) the 1925 Conference on Frost Action sponsored by the Swedish Institute for Roads, 2) Beskow's (5) 1935 treatise, 3) the 1952 review of frost action literature by Johnson (27) covering the period 1905-1951, and 4) the 1974 publication, "Roadway Design in Seasonal Frost Areas" (National Cooperative Highway Research Program) (47). The increased understanding brought about by these contributions went far beyond frost action processes as applied to roads.

The three conditions involving water, temperature and soil that must exist simultaneously for frost heaving to occur in soils are well known, namely a soil moisture supply, sufficiently cold temperatures to cause freezing, and a frost-susceptible soil. The simultaneous occurrence and the interaction process that results is referred to as the "mechanism of frost heaving." Soil type, however, still remains the major factor in defining the susceptibility to frost heaving. Fine-grained soils, including dirty gravel, may exhibit very low to very high heave, depending on the amount of fines, hydraulic conductivity and density of the soil. Dense clays having very low hydraulic conductivity may produce only ice crystals or small ice lenses during freezing. More permeable silts can develop a myriad of ice lenses on freezing, including thick ice layers, attracting additional water to the freezing front from great depths. Conversely, clean sands and gravels may experience low or negligible heave on freezing and in some cases may expel some in-situ pore water during the freezing process.

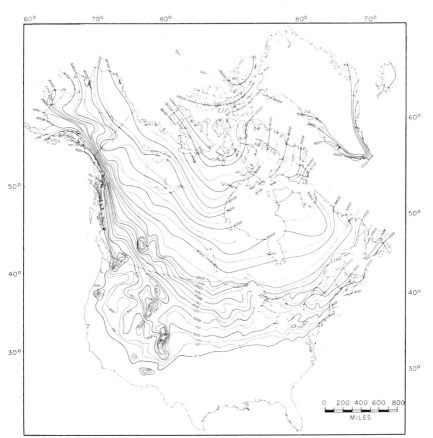

Figure 1. Distribution of Mean Air Freezing Index values (°F) in North America (58)
(Prepared by G.D. Gilman.)

    The frost susceptibility classification of soils is the first step in
assessing the potential frost heave (13, 34, 56).

Extent of Frost Action Problems in North America

    Damage due to frost action is widespread; it occurs in the
temperate zones wherever seasonal soil freezing occurs, as well as in
the active layer of more northerly permafrost regions.

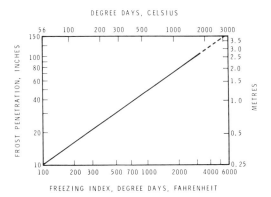

Figure 2.   Relation between Freezing Index
and depth of frost penetration (8).

The depth to which frost penetrates below the surface of the
ground kept clear of snow depends principally on the magnitude and
duration of the below-freezing air temperatures, of which the freezing
index[1] provides a measure. An approximate value of the mean freezing
index may be obtained from Figure 1, which shows the distribution of
mean air-freezing index values in North America. Correlations between
frost depth and the freezing-index (Fig. 2) can be useful as guides to
estimating total frost depth. While freezing indices may vary from
year to year, the southern limit where frost may be generally expected
to penetrate pavements is roughly along the 0°F-day isoline (Fig. 1).
An obvious conclusion is that potential frost heaving problems due to
seasonal freezing become more severe as one proceeds from south to
north or from smaller to larger freezing indices and hence to greater
depths of frost penetration.

FOUNDATIONS (Crory)

Below-freezing air temperatures, while highly variable from day
to day and year to year, penetrate the soil to cause frost heaving and
produce higher short-term strength in the soils when frozen. The
frost penetration depth is rarely the same around a foundation, being
highly influenced by variations in building heat or insulation,
shading, snow cover, and soil conditions, particularly soil moisture.
Fine-grained soils with higher moisture contents, and associated
latent heat, experience less frost penetration than gravels under the
same freezing conditions (28, 34). Because frost penetration into the
ground and frost heaving occur in directions parallel to the direction
of heat flow, the direction of heave may be other than vertical if the
surface is not horizontal. Frost penetration and heaving can be in a
horizontal direction, as in the case of a vertical wall, or at various

---

[1]The "freezing index" is simply the accumulated total of degree-days
of freezing for a given winter.

angles, in the case of merging frost penetration from two directions.
Non-uniform frost penetration also occurs at isolated and unheated
foundation members exposed above the ground surface, such as steel or
concrete transmission tower foundations.

Foundations constructed on frost-susceptible frozen ground can
experience severe settlement and distress when the ground thaws in the
spring. Foundations subjected to frost action during winter construction, when the structure is unheated and not backfilled to final
grade, can experience severe differential heaving. Similarly,
unheated full basements, with only the first floor capped, may suffer
severe wall and floor heaving during the winter. Thus foundation
designs based on the normal operating conditions of a heated basement
must consider the possible consequences of frost heaving that may
occur before such conditions are actually achieved.

Normal and Tangential Frost Heave Forces

The forces on foundations created by heaving of frost-susceptible
soils can be subdivided into two groups -- those normal and those
tangential (3, 28, 34, 56). The normal forces are those at right
angles to the plane of freezing. A continuous strip footing, or
isolated footing, placed on the surface of frost-susceptible material
is therefore subjected to these normal forces, and consequent heaving,
during freezing of the soil directly beneath the footings. When the
footings are placed a foot or more below the ground surface, the
normal forces are not activated until the frost has penetrated below
the base of the footing. If the footing base is placed below the
maximum frost depth, the footing base will not experience any normal
heaving forces, unless the presence of the foundation member locally
accelerates the frost penetration. Should a deep footing, such as an
unheated basement wall or a retaining wall, be exposed to below-
freezing temperatures on one side, there may be unevenly distributed
normal forces on the base of the footing, as well as horizontally
oriented normal forces on the back of the wall. These normal forces
are generally assumed to be only as great as the resisting forces,
such as the weight of the footing, the imposed load, and the surcharge
weight from the overlying soil. While the amount of displacement
caused by frost can be reduced by applied pressures (34), the normal
forces generated by ground freezing usually exceed the conventional
loads imposed on footings. To suppress all frost heaving the footing
loads would have to be in excess of the thawed bearing capacity of
fine-grained soils, since the maximum potential force generated by
such frost heaving stresses can approach that generated by the
confined freezing of water, i.e. greater than 10 tons/ft^2 (9.7
kgf/cm^2) (54, 56). Thus, there are limited options in suppressing the
normal forces and heaving associated with the freezing of frost-
susceptible soils. Such forces and displacements can be avoided,
however, by placing the footing at depths which exceed the maximum
frost penetration, or using non-frost-susceptible soils.

Tangential frost heaving forces are the result of the action of
soils that freeze to the side faces of foundations, the forces being
tangent to such surfaces (24, 31). Tangential forces generated by

frost heaving can best be illustrated by a timber pile which extends
well below the maximum depth of frost, such that it is not subjected
to the normal heaving forces described above. Assuming for discussion
that the pile does not thermally influence the frost penetration, the
soil surrounding the pile will heave upward as it freezes. The soil
will also freeze to the pile. Shearing and tension stresses will be
generated in the frozen soil layer immediately surrounding the pile,
since the normal vertical forces at the bottom of the surrounding
frost front will be thrusting upward, while the frozen soils adhering
to the pole are restrained by the imposed loads on the pile and the
friction along the surface of the pile at greater depth. The
tangential stresses on the frozen layer at the pile surface are
greatest during periods of rapid frost penetration into the surrounding soil. During these same periods the already frozen soils are
coldest and have their greatest strength, as does the associated
adfreeze bonding to the pile. During warm periods, but not necessarily those above freezing, the frost layer is weaker and can bend or
otherwise relax. The adfreeze bond to the pile will also be weaker in
warmer periods, with the creep rate increasing to the extent that the
bond may be broken. When colder weather returns, the tangential
forces increase as soon as the frost begins to penetrate further. In
interior Alaska, the peak tangential heave forces on 8-in. steel pipe
piles have been recorded at more than 55,000 lbf, when the frost depth
was only 2.5 to 3.0 ft (0.7 to 0.9 m) below the ground surface (17).

     Foundation walls are also subjected to tangential heaving forces
(41, 44). When the insides of basement walls are heated, the adfreeze
bond stresses on the outside of the wall will normally be very small
and intermittent. Should the wall be cold, however, such as in an
unheated crawl space or in an unheated garage, the adfreeze bond and
tangential forces may be very high. Distress to foundations can also
occur by a combination of tangential and normal forces acting in
unison, particularly in the latter part of the winter.

     When piles, poles or other foundation members are incapable of
resisting such upward heave forces, the foundations are progressively
heaved upward each year (24, 53). The cumulative heaving over many
years may then be several inches to several feet. The cumulative
frost heaving of a power pole is shown in Figure 3, the braced rigger
being originally at the ground surface.

Design of Foundations for Frost Heave Conditions

     Building codes commonly specify that the bottom surface of
footings, grade beams, pile caps or other foundation construction
shall be below the frost line for the locality, as established by
local records or experience (7, 38, 51, 55). The intent of such codes
is to avoid the normal heave forces, described above. Tangential
heave forces are addressed in recent codes in Canada and Norway which
emphasize an engineering approach to frost heave design, rather than
the simple rule of thumb based solely on frost depth (10, 11, 48).

     The building codes of most cities and towns specify minimum
footing depths which are usually less than the maximum frost depths

Figure 3. Power pole heaved by tangential frost heave
forces in Alaska.

reported by the National Weather Service (21, 50). Frost penetration
depths based on theoretical or empirical equations utilizing long-term
freezing indices and site-specific thermal properties are rarely used
for foundation designs except for ice skating arenas, cold storage
facilities and snow-free pavements. Few details are available on how
local codes established the design frost depths and how they relate to
different foundation types (56). While the minimum depths of footing
placement for frost considerations in the northern states of the U.S.
and in Canada usually range from 3 to 5 ft (0.9 to 1.5 m), the depth
does not increase as one goes north. The city of Fairbanks, Alaska,
building code specifies a minimum foundation depth of only 3.5 ft (1
m) for frost considerations, yet the frost penetration may be more
than 10 ft (3 m) in the Chena River sands and gravels which are
commonly found throughout the city. This deep frost penetration, in
non-frost-susceptible soils, has not caused damage to shallow foundations.

    Foundation distress from frost heaving in the case of fully
heated basements is rarely encountered, provided the basements were
protected during construction and have been heated, or semi-heated
since construction. The greatest distress near heated basements
occurs at outside steps, loading aprons and isolated footings for car
ports or porches, which receive little or no heat from the basement.
The shallow footings of some split-level houses or buildings erected
with unheated crawl spaces are susceptible to frost heave damage, when
founded on or backfilled with frost-susceptible soils. The recent
trend of insulating the interior of foundation walls to conserve heat
can aggravate frost heaving.

Figure 4. Concrete block garage destroyed by normal frost heave forces beneath wall footings.

Accessory buildings, usually limited to one-story structures with less than 400 ft^2 (36 m^2) of floor space, are commonly exempt from the minimum foundation depths specified by National or local codes (7, 38, 51, 55). When such buildings are not continuously heated in the winter, the foundations can suffer differential frost heaving, if built on frost-susceptible soils. Structural damage of such structures may range from minor for wood frame or major in the case of masonry block. The two-car garage shown in Figure 4 was destroyed when the owners thought they could correct the differential frost heaving which had plagued the doorways for years by providing drainage on the sides and rear. Unfortunately the ditch around the garage in this side hill location had little influence on lowering the water table and the greater exposure of the footings allowed much deeper frost penetration. The resulting heave rapidly destroyed the garage. The north (left wall) completely collapsed.

Whenever necessary, and especially in the cases of piles, poles, foundations of unheated structures, and isolated footings, provisions must be included in the design to avoid or resist the displacements associated with the tangential forces of frost heaving (44, 53, 54). In such cases, heave may be avoided by backfilling around the foundation member to sufficient depth and distance with non-frost-susceptible material, by backfilling with a non-heaving low shear strength soil-oil-wax material, or by using oil-wax filled casing which protects the member, as used for frost-free benchmarks in cold regions (3, 34, 54). The tangential forces on piles or poles may also be reduced by the use of coatings, such as coal tar, plastic, Teflon or other materials. The tapering of footings through the seasonally frozen layer and installing timber piles butt down are construction techniques commonly used in cold regions to reduce and counteract tangential frost heaving forces (34). Foundation elements may be designed to resist heave by such techniques as providing adequate depth imbedment in soil or permafrost below the annual frost zone, anchoring foundation elements into bedrock if circumstances permit, or using footing enlargements to develop passive soil resistance. In permafrost areas, thermal piles may be used to increase pile stability, including resistance to heave.

The design of foundations in areas of deep seasonal frost has undergone dramatic changes in the past decade. Many practicing engineers in the U.S. are unaware of these changes, however, because most of the code changes have occurred in other countries. Norwegian building codes no longer specify foundation depths based on frost depth. The foundations must be designed to prevent damage from frost action on a performance concept. Designs for basements, crawl spaces and slab-on-grade construction in Norway are based on anticipated heat flow patterns and frost susceptibility, taking full advantage of the use of insulation under and around foundations (1, 24, 32, 41, 48, 52). Canadian building codes and standards (10, 11) also stress the importance of engineering the design of foundations in frost areas, rather than placing the burden of design on minimum depths of embedment. Soviet, Canadian and Norwegian design requirements include the consideration of tangential heaving forces.

DESIGN OF PIPELINES FOR FROST HEAVE (Isaacs)

Until very recently, no rational system has been available for frost heave design of buried pipelines[2]. However, the following material gives a quick overview of a recently developed approach to this problem, aimed primarily at the case of pipelines operated at below-freezing temperatures.

In the transportation of natural gas from the Far North, disturbance of the ground surface due to construction activities will affect the existing permafrost, possibly causing thaw, settlement and in some cases slope instability. One solution to mitigating these undesirable

---

[2]Editorial footnote: Frost heave of pipelines carried on piles or other structural supports may be controlled by application of the approaches discussed under Foundations, above.

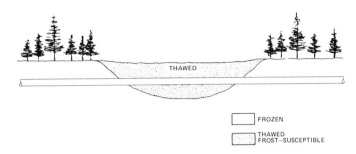

| | FROZEN |
| | THAWED
FROST–SUSCEPTIBLE |

Figure 5. Buried, chilled gas pipeline in frozen and thawed soils.

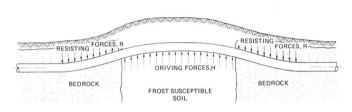

Figure 6. Heave span at soil transition.

effects is the transportation of the gas at temperatures less than
32°F.

This solution causes an additional problem, however. A frost
bulb is generated around the pipe where it passes through unfrozen
soils (Fig. 5). As the pore water within these soils freezes and as
additional water is attracted to the freezing front in some of these
soils, heave develops, tending to lift the pipeline. When the pipeline passes through soils with different heave properties, differential heave results (Fig. 6).

The major design problems then faced are prediction of 1) the
heaving or driving forces, H, 2) the resisting forces, R, and 3) the
span over which differential heave occurs (Fig. 6).

To assess how much soils will heave and under what conditions,
current practice involves testing soil samples under very carefully
controlled laboratory conditions. Samples are placed in a rubber
membrane, in a manner similar to that used in the triaxial testing in
soil mechanics, but are contained within a Perspex cell. Between the
cell and membrane a lubricant is used to limit side friction. In the
walls of the cell are thermistors at 0.25-in. (6-mm) spacing. Four
inches (10 cm) of urethane insulation surround the cell, and temperatures at the top and bottom of the sample are controlled by the
circulation of ethylene glycol in aluminum plates (see Fig. 7). The
equipment is kept in a coldroom maintained at approximately 36°F
(2°C). A water supply is provided at the top of the sample.

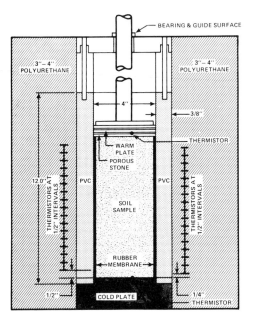

Figure 7.   Typical frost heave test cell.

The laboratory test consists of the following phases:

1. Consolidation of the sample within the cell at the test
    pressure.

2. Back-pressuring of the sample to ensure saturation with
    water.

3. Tempering of the sample to the warm (top) plate temperature.

4. Nucleation of the sample at the cold (base) plate.

5. Linear reduction with time of the top and bottom plate
    temperatures to control the rate of frost penetration.

    In these tests, the surcharge pressure, thermal gradient and
frost penetration rates are imposed on the sample with the intention
of duplicating as closely as possible the thermal and stress histories
to which a sample at a particular depth beneath the pipe operating at
a specified temperature will be subjected. Heave, water intake, and
frost penetration as a function of time are the measured parameters.
Other standard soils data such as density, gradation, and so forth are
also collected.

To predict heave, proprietary finite-element heat transfer models are used to calculate the growth of the frost bulb with time. With the appropriate laboratory data input, heave with time is predicted and compared with the results being obtained from a number of field test sites in Alaska.

If a section of pipeline heaves more than an adjoining section, the soil and/or frost bulb above the pipe in the lesser heaving section will exert a force resisting the displacement of the pipe. This is referred to as the uplift resistance. The value of this resistance will vary with the time of year since the thermal state of the soil above the pipe varies with the seasons. It will also vary with the burial depth, the rate at which the pipe moves, the type of soil, the dry density and the moisture content of the soil.

Derivation of the uplift resistance relationship is currently based on the results of small-scale laboratory and full-scale field tests where a pipe is moved at constant rates of displacement through frozen soil. Laboratory data on unconfined compression and tension tests on frozen samples at different temperatures and strain rates are also being employed in a finite-element computer program to predict the variation of resistance with soil type, density, temperature and rate of displacement.

For the design of the pipeline, accurate logging of the landforms and their associated features along the route is essential. To delineate prior to construction the infinite number of spans over which differential heave would occur in various soil conditions would, however, be unreasonable and imprudent. To decide in the field during construction would create unacceptable consequences in costs and delays. The design is therefore based on the development of a critical span for each landform and designing the pipeline for that critical span.

Where soils exhibit a heave relationship which is strongly pressure-dependent, a curve of pipe strain vs pipe span can be developed which exhibits a maximum strain at a particular span called the "critical span." The strain at the critical span will then be compared to the maximum allowable pipe strain (Fig. 8).

Where calculations indicate that the strain would eventually reach the allowable strain limit or where the soils exhibit large heaving tendencies and a limited dependency on pressure, mitigative measures would be used during construction. For unforeseen conditions, remedial action would be taken some time after completion of construction but prior to the time when the strain limit is reached.

If the heave pressure and uplift functions for a landform are such that the maximum strain at the critical span of the pipe, within its lifetime, will not exceed the maximum allowable strain, no corrective measures would be required.

To develop the critical span relationships as well as to study other aspects of pipe-strain behavior, additional proprietary computer

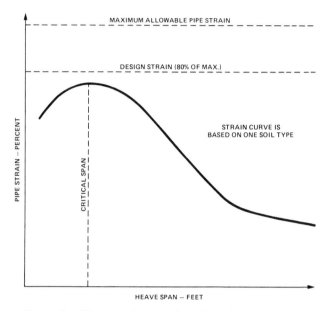

Figure 8. Pipe strain as a function of heave span.

programs are used. The programs are two-dimensional finite element
models of a beam on an elastic foundation using Winkler springs to
simulate the force-displacement relationships.

     The above design procedure for a pipeline to resist frost heave
is a logical process. It is based on the collection of high quality
data from properly structured laboratory tests, the employment of
these data in sound computer programs to predict field behavior, and
the verification of these predictions from the results of closely
controlled field experiments.

FROST HEAVE DUE TO ARTIFICIAL FREEZING AND TO IN-GROUND STORAGE OF
CRYOGENIC LIQUIDS (Sanger)

Artificial Ground Freezing

     Soil displacements under artificial freezing are rarely
published, although routine measurements for both vertical and
horizontal heave are generally made during and after freezing.
Unfortunately when heave data are given, they are seldom accompanied
by an adequate description of soils or other conditions. For most
soils frozen artificially, the limiting effective pressure is from 150
to 1500 psf (4 to 40 kN/m^2). With vertical freeze-pipes the ground
heave is commonly from 4 to 6 in. (10 to 15 cm) and the lateral heave
is from 3 to 6 times that, due to pore water phase-change. There
seems to be no way of predicting lateral heave except by experience

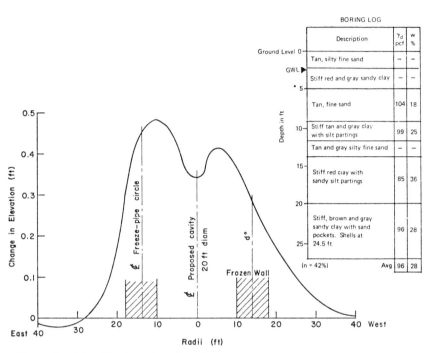

BORING LOG

Description	γ_d pcf	w %
Tan, silty fine sand	–	–
Stiff red and gray sandy clay	–	–
Tan, fine sand	104	18
Stiff tan and gray clay with silt partings	99	25
Tan and gray silty fine sand	–	–
Stiff red clay with sandy silt partings	85	36
Stiff, brown and gray sandy clay with sand pockets. Shells at 24.5 ft	96	28
(n = 42%) Avg	96	28

Figure 9.   Distribution of maximum heave at the Lake Charles pilot project for inground storage of liquified methane after 64 days of freezing (46).

with similar soils and conditions. Heave is reduced by increasing the rate of freezing, using lower temperatures (commonly freeze pipe temperatures are between -4°F and -40°F [-20°C and -40°C]); the ultimate method of increasing the freezing rate is to use liquid nitrogen at about -180°C. Heave by in-situ freezing can be approximated (49), although it involves a factor of ignorance. Figure 9 shows the distribution of heave at the Lake Charles Project (46) for ground storage of liquefied methane; the ground was prefrozen and the data are for the first 64 days of freezing, at -49°F (-45°C). The heave outside the frozen wall is usual and thought to be attributable to upward flow of soil under lateral pressure. The subsidence of the ground surface at radial distances greater than 15 ft (3 m) from the frozen wall is similar to that reported by Collins and Deacon (15). In a tunnel project at Montreal (35) a heave of 5 in. (12.7 cm) was recorded. The rate of heave can be quite variable; at Lake Charles it was about 0.09 in. (2.0 mm) a day. Khakimov (30) gives limited but good data on heave and rates of heave. Further data are to be found in Endo (20) in his paper on the Kanasugi Bridge crossing of the subway to Tokyo.

Khakimov (30) tried to compute the lateral pressure set up by
freezing but the results, based on ideal elasticity, are doubtful in
soils, unfrozen or frozen. Chen et al. (14) measured pressure on
shafts in freezing soils: 3500 psf (1.7 kgf/cm^2) in sandy gravel and
5700 psf (2.8 kgf/cm^2) in silty clay. Pressures in artificial
freezing are unlikely to crush freeze-pipes but heave and differential
lateral heave can cause breakage when the pipes penetrate layers of
differing soils, especially at welds. Some contractors employ used
pipe for freezing to save cost. New pipe is recommended, except for
sands. Freeze pipes tend to fail at welds. In addition to temperature measurements and probing to check on the progress of the freezing
front, observations of heave and other phenomena are important.
Miyoshi et al. (37) and Heinrich et al. (23) are good references for
modern instrumentation when artificially freezing ground.

Heave is likely to be significant for structures, especially
since it is often non-uniform, so it is inadvisable to build on or
near ground that is to be frozen. Structures on or near the site can
be protected by a circulating warm-water system, similar to that used
for freezing, or possibly by jacks and wedges. Ground subsidence
while thawing the frozen ground is a danger which is often overlooked. The report on the Sao Paulo underpinning project (18) is
worthy of careful study, showing the interaction between structure and
soil-freezing; the remolding of sands was astonishing.

Frozen Ground Storage for Cryogenic Liquids

The techniques for in-ground storage of cryogenic liquids were
developed by Conch International Methane Ltd. of London in the late
1950's. To date only liquid natural gas (LNG) and liquid petroleum
gases (LPG) have been stored in that way (26, 36, 46). Natural gas
cannot be liquefied at any pressure, at atmospheric temperatures.
Storage at a pressure slightly above atmospheric requires refrigeration to -260°F (-162°C). LPG is normally stored at -45°F (-43°C).
Following the pilot project at Lake Charles in 1961 (46) the first
commercial storage facility was built at Arzew, Algeria (26) for
200,000 bbl (38,000 m^3) of LNG. The pit is 122 ft (40 m) in diameter
and 114 ft (38 m) deep; artificial freezing was used for excavation
and the initial cool-down. Soon afterwards a large storage facility
was built on Staten Island, New York, but it had heating coils under
the floor and was in principle an above-ground tank buried in a fill.
At about the same time a smaller LPG storage was built, also by pre-
freezing, at Woods Cross, Illinois (36).

The Arzew storage was in sand-fill, tuff and variable marls with
water-bearing seams. Close to the tank the heave was less than 1 ft
(30 cm), which did not increase with time after filling with LNG. The
heave gradually spread outwards with time but was of a smaller amount,
and possible damage to pipe racks nearby was prevented by the use of
jacks and wedges. As with temporary ground freezing, a structure can
be protected by a wall of warmed soil, if lateral heave is not serious
at distance from the tank. Cryogenic in-ground storage is still under
development in Japan, Belgium and perhaps in other places.
Inexperience of engineers and contractors (especially their workmen)

has been a continuous problem. Training and experience, plus
knowledgeable supervision, are of the utmost importance in both
artificial ground freezing and in cryogenic in-ground storage.

PAVEMENTS AND OTHER ROAD BEDS (Shook)

Considerable literature has been devoted to the destructive
effects of frost heave on pavements and other road beds during the
past 25 years. The following paragraphs cover very briefly some of
the information available. However, the references cited should be
consulted for more extensive treatment of the different subjects
covered, including frost heave mechanisms, general design recommendations, examples of design practices in the United States and other
countries, the use of insulating layers to retard frost penetration
and the use of stabilized bases in a frost heave environment. Papers
by Aldous (2), Argue, (4) Berg (5), Caniard and Peyronne (12), Croney
(16), Linell (33), Nordal (39), and others (40), (43) and (45) are
recommended.

Major distress caused by frost heaving soils on pavements and
other roadbeds often is manifested by dislocation of culvert head
walls, with bumps and rises caused by rocks and cross-drainage pipes
being raised by frost action. Rough roadbeds caused by non-uniform
soil conditions and localized failures from frost boils or other weak
areas are additional problems. In highway pavements, wide cracks down
the center of the pavement or smaller edge cracks often can be traced
to non-uniform transverse conditions. Frost distress increases
maintenance costs and frequently requires costly repairs in locations
where access to good quality agrregates is limited, or where proper
maintenance equipment is not available. Distress from frost heave is
increased by heavy wheel loads, particularly during thaw periods.

A number of theories have been proposed to explain frost heaving,
which are discussed in more detail elsewhere (13), and these concepts
can be useful in assessing the value of proposed methods for minimizing the effects of frost heave on roadbeds. In general, the more
important conditions or characteristics involved in heaving are 1) the
rate of heat removal, 2) the weight of materials supported by the
frozen soil (or the magnitude of the overburden pressure), and 3)
various physical characteristics of the soil or the soil/water
combination. Controlling these conditions will help minimize the
effects of frost heave.

Proper roadway design for areas where frost-susceptible soils are
known to exist, or are suspected, requires that the roadway site be
thoroughly investigated. The type and extent of susceptible soils
need to be determined, potential drainage problems located and a
measure of potential frost penetration determined (4, 13, 16, 29, 42).

Non-uniform or differential frost heave appears to be the most
destructive characteristic of frost heaving. Any roadbed design that
produces a uniform subgrade condition will help minimize damage where
conditions for frost heaving exist. Non-uniform heave occurs
frequently in transition sections from cut to fill areas, but also
occurs because of the variable nature of existing soil strata. Mixing

existing frost-susceptible and non-frost-susceptible subgrade soils
has been recommended, but this can be expensive. A less expensive
technique is to remove and replace small areas of soils highly susceptible to frost heave. Gradual transition sections from cut to fill
areas can be designed so that abrupt changes from non-heaving to
heaving soils do not occur.

Special attention to drainage is necessary. Surface drainage
must prevent entry of water from a pavement surface. Natural discharge or drainage installed during construction must be designed and
maintained to prevent blockage by snow or accumulated debris.
Lowering the water table may be possible in some situations. Changes
in the water table brought about by cutting through existing ground
features must be anticipated and designed for.

One of the more common methods for minimizing frost heave is to
remove and replace up to several feet of the frost heaving soil with
non-frost-susceptible material. Although such techniques do not
necessarily reduce the depth of frost penetration (25), they do
provide added overburden pressure, improve drainage for water that
collects during the thawing phase of the process and, probably as
important as anything, help provide uniform support throughout the
length of the pavement. Non-uniformly shaped transverse subbase cross
sections have been used to minimize longitudinal cracking (2, 39, 40).

A common requirement is that non-frost-susceptible layers be
substituted to a depth of 50 percent of the anticipated depth of frost
penetration, although, depths of over 75 percent have been specified
(33, 47). Depth of frost-free material required also may vary with
the type of pavement and class of highway. One method for selecting
base thickness to limit frost penetration is illustrated by Figure 10
(5). The intention is to limit frost penetration into frost-susceptible soils to such a level that any differential heaving will not
cause excessive surface roughness for high speed traffic or unacceptable cracking of the pavement surface.

The use of stabilized pavement base layers has been recommended
for reducing the need for thick granular layers for frost penetration
(12, 19, 24, 45). Experiments indicate that in some cases frost depth
penetration is retarded by stabilized layers, and, while heave does
occur, it is minimized. If this approach is considered, it is
especially important that drainage and possible non-uniform heaving
conditions are given special attention.

Depths of frost penetration or rates of heat removal can be
effectively controlled by the use of various materials between the
frost-susceptible soil and the pavement structure. These include both
natural materials, such as wet sand, peat or bark, and in more recent
years such man-made insulating materials as Styrofoam and foamed
sulphur (9, 22, 39). Placing insulating layers normally requires
considerable hand labor. Man-made insulation is expensive and sometimes promotes the formation of ice on the pavement surface, creating
dangerous driving conditions.

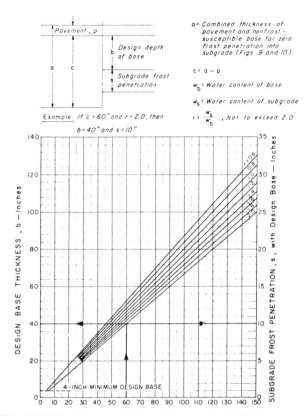

Figure 10. Design depth of non-frost-susceptible base for
limited subgrade frost penetration.

As indicated above, a number of methods are available for minimizing detrimental effects of frost heave on roadways. Unfortunately,
not all are completely successful and most are expensive. Good subsurface investigation during the design stage, coupled with selective
application of different design techniques, should produce the best
results, consistent with available resources.

CONCLUSIONS ON THE STATE OF THE ART AND RESEARCH NEEDS

Foundations

The current practice of simply specifying a minimum depth of
footings based on a locally determined depth of frost is considered
inadequate and completely ignores the effect of tangential frost heave

forces. The U.S. codes should be revised in light of the more
extensive performance-related criteria of the U.S.S.R., Canada and
Norway. An extensive educational program for designers, builders,
inspectors, and owners will be required to facilitate changes in
existing codes. Greater emphasis must be placed on foundations
designed by geotechnical engineers, competent in cold regions work.
Research and technology transfer will be required in developing more
precise frost susceptibility criteria, a more rapid frost susceptibility test, and heat transfer analysis techniques applicable to a
wide range of foundation types.

Pipelines

In developing the criteria for the design of the chilled gas
pipeline in Alaska, only limited studies have been made with respect
to the heaving behavior and uplift resistance of clay-type soils,
since they do not occur along the current right-of-way. In other
areas, however, research into these soils may be necessary, especially
as it is suspected that clays will behave very differently from silty
soils.

While attempts are being made to incorporate convection within
existing thermal models, it is difficult to simulate and even more
difficult to verify. An example of the importance of this modeling is
the case of a chilled pipeline crossing beneath a stream where water
flow through the bed material might be partially or completely blocked
by the frozen zone surrounding the pipeline.

Long-term climatic and ground temperature data are necessary for
accurate simulations of thermal behavior. Reliable and accurate
recording equipment, capable of operating unattended at remote sites
in such hostile environments, does not appear to exist, as yet.

Where they may be a requirement for the use of insulation on
below-ground pipes, existing data on the reliability of such insulation over a long period of time and over a number of freeze-thaw
cycles are inadequate.

Artificial Ground Freezing and In-ground Storage of Cryogenic Liquids

More field frost heave data should be obtained, and published, on
ground freezing. Radial and axial heave should be correlated with
soil and freezing conditions. Subsidence on thawing after construction also needs additional investigation. The risks of fissuring in
some soils upon thawing, particularly overconsolidated clays, should
be studied. Important topics for research are the rate of growth,
shape and spread of the frost bulb with time, effects of surface and
lateral heave, and fissuring. In-ground storage problems which should
be resolved included the requirements for liners which deform
elastically and the interaction of such liners with plastic frozen
soils which creep.

Pavements and Other Roadbeds

Frost heave can cause serious problems with highway pavements and
other roadbeds. Engineers responsible for roadway design need to
thoroughly investigate areas where potential frost heaving soils may
occur, and to apply design techniques on a selective basis for the
most economical results. Unfortunately, frost heave cannot be completely eliminated in most cases. The literature available indicates
that many recommendations for minimizing detrimental effects of frost
heave were developed over 20 years ago, and this may be a fruitful
area for future research.

ACKNOWLEDGMENTS

The various sections of this paper have been prepared by the
following individuals: Introduction (combined), E. Penner; Founda-
tions, F.E. Crory; Design of Pipelines for Frost Heave, R.M. Isaacs;
Frost Heave due to Artificial Freezing and to In-ground Storage of
Cryogenic Liquids, F.J. Sanger; Pavements and Other Roadbeds, J.F.
Shook; and Conclusions on the State of the Art and Research Needs,
jointly. The authors thank Kenneth A. Linell for coordinating the
preparation of this report on behalf of the Committee on Freezing and
Thawing of Soil-Water Systems of the Technical Council on Cold Regions
Engineering.

The design approach for buried pipelines outlined in this paper,
including the laboratory and computer techniques, has evolved over a
number of years and is the result of contributions of numerous
individuals. However, the following have made significant contributions to the Northwest Alaskan Pipeline (NWA) approach, in particular:
Messrs. J.E. Myrick and C.B. Hazen of NWA, Mr. M. Losey of Fluor
Engineers and Constructors, Dr. K. Meyer of the Joint-Venture for the
project, and Messrs. C. Burrows and R. Rein of R&M Consultants.

APPENDIX - REFERENCES

1. Adamson, B., "Foundation with Crawl Spaces. Frost Penetration
    and Equivalent U-value of Floor Slab," Frost i Jord, No. 8, Oslo,
    1972.

2. Aldous, W.M., "Engineering of Frost Susceptible Soils," Rural
    Roads, August, 1961.

3. Andersland, O.B. and Anderson, D.M., "Geotechnical Engineering
    for Cold Regions," McGraw-Hill, New York, 1978.

4. Argue, C.H., "Frost and Thaw Penetrations of Soils and Pavements
    at Canadian Airports," Proceedings, Canadian Good Roads
    Association, Ottawa, 1969.

5. Berg, R.L., "Design of Civil Airfield Pavements for Seasonal
    Frost and Permafrost Conditions," Report No. FAA-RD-7430, Federal
    Aviation Administration, Washington, D.C., October, 1974.

6. Beskow, B., "Soil Freezing and Frost Heaving with Special
    Application to Roads and Railways," Sver. Geol. Unders., ser. C,
    vol. 375, 1935, trans. J.O. Osterberg, Technical Institute,
    Northwestern Univ., Evanston, Ill., 145 pages. 1947.

7. BOCA Basic Building Code/1978, Building Officials and Code
    Administrators International, Inc., Homewood, Ill, 1978.

8. Brown, W.G., "Difficulties Associated with Predicting Depths of
    Freeze and Thaw," Canadian Geotechnical Journal, vol. 1, no. 4,
    1964, pp. 215-226.

9. Campbell, R.W., Woo, G.L., Antoniades, E.P. and Ankers, J.W.,
    "Sulfur Foam: Sudic-Chevron Field Test for Frost-Heave
    Prevention," Proceedings, Canadian Technical Asphalt Association,
    Victoria, 1975.

10. Canada National Research Council, National Building Code of
    Canada. Associate Committee on the National Building Code,
    Ottawa, Ont., 1980.

11. Canada National Research Council, Supplement to the National Code
    of Canada, Associate Committee on the National Building Code,
    Ottawa, Ont., 1980.

12. Caniard, L. and Peyronne, C., "The Consideration of Frost in the
    Design of Asphalt Pavements," Proceedings, Fourth International
    Conference on Structural Design of Asphalt Pavements, Ann Arbor,
    Michigan, 1977.

13. Chamberlain, E.J., "Comparative Evaluation of Frost-Suscepti-
     bility Tests," Transportation Research Record 809, Washington,
     D.C., 1981.

14. Chen, Hsiao-pai and Wu, Tzu-wang, (1978) "Experimental Research
     on the Principal Mechanical Properties of Freezing and Frozen
     Soils in China," Proceedings, Third International Conference on
     Permafrost, vol. 2, NRC, Canada.

15. Collins, S.P. and Deacon, W.G., (1972) "Shaft-sinking by
     Ground-freezing; Ely Ouse-Essex Scheme," Proceedings, Inst. Civ.
     Eng., Paper 7506S, (1972 Supplement).

16. Croney, D., "Damage to Roads Caused by the Frost of 1962-63,"
     Laboratory Note No. Ln/459/DC, Road Research Laboratory, Great
     Britain, November, 1963.

17. Crory, F.E. and Reed, R.E., "Measurement of Frost Heave Forces on
     Piles," U.S. Army Corps of Engineers, Cold Regions Research and
     Engineering Laboratory, Hanover, NH, Technical Report 145, 1965.

18. Dumont-Villares, A., "The Underpinning of the 26-Story 'Companhia
     Paulista de Securos' Building, Sao Paulo, Brazil," Geotechnique,
     ICE, London, vol. vi, Nov. 1956.

19. Eaton, R.A., and Joubert, R.H., "Full-Depth Pavement
     Considerations in Seasonal Frost Areas," Proceedings, Association
     of Asphalt Paving Technologies, 1979.

20. Endo, K., "Artificial Soil-Freezing Method for Subway
     Construction," Civ. Eng. in Japan, Soc. Civ. Eng., Tokyo, 1969.

21. Fletcher, G.A. and Smoots, V.A., Construction Guide for Soils and
     Foundations, John Wiley and Sons, New York, 1974.

22. Gandahl, R., "Plastic Foam Insulation of Roads," National Road
     and Traffic Research Institute, Sweden, 1981.

23. Heinrich, D., Muller, G., and Voort, J., "Ground-Freezing
     Monitoring Techniques," Proceedings, International Conference on
     Ground-Freezing, Univ. of Bochum, Germany, 1978.

24. Herje, J.R., "Pillars and Piles in the Ground: Frost Problems,"
     Frost i Jord, No. 8, Oslo, 1972.

25. Hinderman, W.L., "The Swing to Full-Depth," Information Series
     No. 146 (IS-146), The Asphalt Institute, June, 1968.

26. Jackson, R.G., "Ground Storage at the CAMEL Gas Liquefaction
     Plant at Arzew," Europe and Oil, January, 1966.

27. Johnson, A.W. "Frost Action in Roads and Airfields, a Review of
     the Literature," Special Report No. 1, Highway Research Board,
     Washington, D.C., 1952, 287 pp.

28. Johnston, G.H., "Permafrost-Engineering Design and Construction,"
     John Wiley and Sons, Toronto, 1981.

29. Jumikis, A.R., Thermal Soil Mechanics, Rutgers University Press,
     New Brunswick, New Jersey, 1966.

30. Khakimov, Kh.R., "Artificial Freezing of Soils; Theory and
     Practice." Trans. from the Russian by A. Barouch, Jerusalem,
     Israel Program for Scientific Translations, 1966. Available from
     U.S. National Techincal Information Service (NTIS), Springfield,
     VA 22151.

31. Kinoshita, S. and Ono, T., "Heaving Forces of Frozen Ground,"
     National Research Council of Canada, Ottawa, Technical
     Translation 1246, 1966.

32. Klove, K. and Thue, J.V., "Slab-on-Ground Foundation," Frost i
     Jord, No. 8, Oslo, 1972.

33. Linell, K.A., "Pavement Design in Frost Areas," presented at the
     Asphalt Paving Conference, Syracuse University, Syracuse, NY,
     March 1960.

34. Linell, K.A. and Lobacz, E.F., eds., "Design and Construction of
     Foundations in Areas of Deep Seasonal Frost and Permafrost,"
     U.S. Army Corps of Engineers, Cold Regions Research and
     Engineering Laboratory, Hanover, NH, Special Report 80-34, August
     1980.

35. Low, G.W., "Soil Freezing to Reconstruct a Railway Tunnel,"
     Journal of the Construction Division, ASCE, No. CO3, Nov. 1966.

36. Massey, P.S., "Frozen Earth Propane Storage," Oil and Gas
     Journal, March 16, 1964.

37. Miyoshi, M., Tsumoto, T., and Kiriyama, S., "Large Scale Freezing
     Work for Subway Construction in Japan," Proceedings,
     International Conference on Ground-Freezing, Univ. of Bochum,
     Germany, 1978.

38. National Building Code, Engineering and Safety Service, N.Y.,
     1976.

39. Nordal, R.S., "Frost Problems in Highway Design in Norway,"
     Public Works, July, 1967.

40. Nordic Cooperative Research Report, "Failure Models and Pavement
     Design and Rehabilitation Systems Developed and Adapted for
     Conditions Prevailing in the Nordic Countries," Proceedings,
     Fourth International Conference on Structural Design of Asphalt
     Pavements, Ann Arbor, Michigan, 1977.

41. Norgard, L., "Frost Problems in Basement Constructions," Frost i
     Jord, No. 8, Oslo, 1972.

DESIGN IMPLICATIONS OF SUBSOIL THAWING

by

Thaddeus C. Johnson, M. ASCE,[1], Edward C. McRoberts[2] and John F. Nixon[3]

ABSTRACT

This paper reviews the state of the art in the geotechnical design of earthen and earth-supported structures affected by subsoil thawing. The analysis of the ground thermal regime during the thawing phase is first summarized. After a brief review of the effects of conditions prevailing during freezing upon the behavior of thawing soils, the pertinent mechanical properties of thawed soils are described in some detail. The paper then reviews the general approaches for geothermal design. The principal focus of the paper is a review of the prevailing methods for geotechnical design for subsoil thawing. Design approaches and examples are given for foundations, work pads, roads and airfield pavements, pipelines, well casings, slope stability and embankment dams. More than one hundred technical publications are cited.

INTRODUCTION

Geotechnical engineering practice in northern North America and Eurasia has long included recognition that thawing of frozen soil adversely affects its compressibility, shear strength and deformability, and may result in large settlements of supported structures, slope failures and pavement distress. Empirical design criteria developed for such facilities as roads, airfields, and building foundations in seasonal frost and permafrost areas served well the identified needs.

With the acceleration of development activities in the North within the past 1-1/2 decades, mainly related to discovery, production and transport of oil and gas, new needs were identified and research and engineering to meet those needs received greater support. Consequently, our understanding of the mechanics of thawing soils, and our capacity to deal with engineering problems in a systematic way, have advanced significantly.

The phenomena controlling the mechanical properties of thawing soils are exceedingly complex. The properties of most critical interest are those relating to shear failure and to deformations originating in either shear strains or volume changes. The state of stress in pore water exerts a governing influence on all three of these distress manifestations.

―――――――――

[1] U.S. Army Cold Regions Research and Engineering Laboratory, Hanover, N.H.
[2] Hardy Associates (1978) Ltd., Edmonton, Alberta
[3] Hardy Associates (1978) Ltd., Calgary, Alberta

Excess pore pressures generated in thawing soils sharply reduce their stability, bearing capacity and shear strength, and consolidation and dissipation of the pore pressures cause large settlements. Consequently for many structures in northern regions, the periods of subsoil thawing represent the most critical conditions affecting the preservation of the integrity of supported structures. For certain types of structures in permafrost areas, such as generating stations and hot oil pipelines, thawing may be continuous and progressive, while in seasonal frost areas thawing may be limited to only a few weeks each spring. In either case the most critically affected soil is that at the thawing front, but of accompanying geotechnical interest is the rate and extent to which the recovery process restores the soil to a condition comparable to the prefrozen state.

This paper deals with the ground thermal regime, particularly during the thawing phase, the effects of conditions prevailing during freezing upon the behavior of thawing soils, and the mechanical properties of thawed soils. The principal focus of the paper is upon techniques and examples of geotechnical designs for subsoil thawing.

GROUND THERMAL REGIME

A primary feature of any design for thawing soils is a geothermal analysis to obtain the depth and rate of thaw, together with information on freezeback, if freeze-thaw cycling is of concern. This thaw and freezeback information is more important than knowing explicitly the temperatures at all points in the ground, although knowledge of the temperatures in the frozen ground will be necessary in permafrost foundation design. The depth of thaw is important in predicting the extent of thermal disturbance and in calculating settlement, for example, whereas the time rate of thaw is important in considerations of excess pore pressure buildup in thawing soils.

Analysis

The complexity of thermal analysis for determining the depth of thaw can vary from simple hand calculations to rather complex numerical models accounting for a wide variety of variable thermal properties and surface effects. Most subsoil thawing problems require relatively minimal effort to obtain the depth of thaw with time. A good review of simple analytical solutions is given by Harlan and Nixon (35). The use of a finite element model to solve a wide variety of permafrost-related problems in heat transfer has been described by Jahns et al. (36). A versatile geothermal simulator using a finite difference method to solve problems involving convection and conduction in radial or rectangular coordinates is described by Nixon and Halliwell (93). Some important solutions for thaw depth are briefly introduced herein.

When a frozen soil of uniform properties is subjected to a sudden "step" increase in surface temperature, the depth of thaw, X, is related to the elapsed time, t, by

$$X = \alpha \sqrt{t} \qquad (1)$$

where α = a constant = $\sqrt{(2k_uT_s/L)}$

k_u = the thermal conductivity of thawed soil

T_s = the surface temperature

L = the latent heat of the soil.

This relationship, known as the Stefan solution for the depth of thaw or freezing in a uniform soil, predicts that the thaw depth will advance proportionally to the square root of time. The frozen soil is taken to be initially at thawing temperature, the properties of the frozen and thawed soil are assumed to be independent of temperature, and the sensible heat of the thawed soil is neglected. Other variations, extensions and improvements have been applied to this solution, and some of these are warranted depending on the particular application. When the effects of surface layers of insulation, peat or significantly different soil properties are considered, a solution for thaw depth in a two-layer system becomes important. Assuming a linear temperature distribution in each layer, and taking the melting temperature as 0°C, the thaw penetration into the lower layer is given as:

$$X - H = \sqrt{(\frac{k_2}{k_1}H)^2 + \frac{2k_2T_s(t-t_o)}{L_2}} - \frac{k_2}{k_1}H \qquad (2)$$

where H = the thickness of the surface layer

t^0 = the time to thaw the surface layer

k^1 and k^2 = the thermal conductivities of the surface layer and the lower thawed layer

L^2 = the latent heat of the lower layer.

For thawing beneath exposed soil surfaces, the surface temperature, T_s, is not equal to the air temperature. The concept of an 'n' factor has been introduced to relate the two, and typically n is about 1.5 for gravel or mineral soil surfaces in summer and about 0.9 for snow-covered winter surfaces. A review of typical n-factors is given by Harlan and Nixon (35). In contrast to the more rigorous method of prediction using a complete surface energy balance, the n-factor approach is much more simplified but yields solutions of sufficient accuracy for many practical problems.

Another particularly useful solution may be extracted from the literature on heat transfer around a buried pipe. This solution is strictly valid only for a pipe buried at great depth, but in the absence of more complex numerical solutions, it provides a reasonable estimate of thaw around a heated pipe buried in permafrost. The solution for the normalized thaw radius, $\bar{R} = R/a$, has the form

$$2\bar{R}^2\ln(\bar{R}) - \bar{R}^2 + 1 = \frac{4k_uT_st}{a^2L} \qquad (3)$$

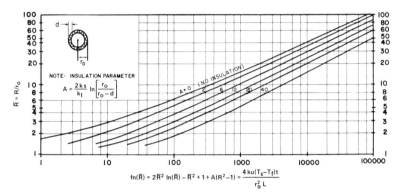

$$fn(\bar{R}) = 2\bar{R}^2 \ln(\bar{R}) - \bar{R}^2 + 1 + A(R^2-1) = \frac{4ku(T_s-T_f)t}{r_0^2 L}$$

Figure 1.   Radial Stefan solution for insulated pipe.

where a = pipe radius

T_s = pipe temperature in degrees above freezing.

The dimensionless thaw radius, $\bar{R}$, is plotted in Figure 1 against the right-hand side of the radial thaw equation. This relationship has been further extended by the authors to account for insulation wrapped around the pipe, and a family of curves has been given on Figure 1 to allow the calculation of thaw depth for different insulation thicknesses. The parameters involved in the solution are defined on the figure.

Armed with these simplified but powerful geothermal analyses for thaw, together with other published charts and solutions for thawing adjacent to warm structures on frozen ground, the designer can proceed to the more challenging aspects of design for thawing subsoil conditions.

Monitoring

The monitoring of the progress of thawing is normally carried out by (a) probing or, in some cases, drilling and retrieval of undisturbed cores, (b) installing strings of temperature-sensing devices known as thermistors, or (c) using geophysical techniques, appropriately calibrated, or "ground-truthed."

Control

The control of the ground thermal regime is a challenging, and sometimes frustrating, task. Except for one notable exception, the following methods of ground thermal control usually involve manipulating one of the components of the surface energy balance, or the near-surface thermal properties:

(a) Placement of natural insulation such as wood chips, sawdust or peat moss in a layer on the surface. These materials will usually absorb water and biodegrade, eventually becoming ineffective, but they also are useful for establishing vegetation on non-traveled surfaces.

(b) Placement of thin layers of sand or gravel fill. These thin layers have little thermal benefit, and may even accelerate thaw due to the low moisture content, different albedo and evapotranspiration properties of the granular fill.

(c) Installation of synthetic insulation layers. This method will greatly retard the rate of thaw, but not completely stop it in warm discontinuous permafrost areas.

(d) Painting of paved surfaces. The rate of thaw can be accelerated by painting a paved surface black, or greatly reduced by painting it white.

(e) Use of polyethylene sheeting. Recent studies by Esch (29) in Fairbanks indicate that surface polyethylene sheeting can increase thaw depth by increasing the "greenhouse" effect in the surface radiation balance.

(f) Removal of vegetation. Long-term observations reported by Linell (56) show that the thaw rate is greatly increased beneath surfaces that are continually stripped or cleared, and it is increased somewhat less in cleared areas that are allowed to naturally revegetate.

(g) Surface heating by hot water or electrical methods. This technique has been occasionally used to accelerate thawing and removal of icy permafrost layers.

Manipulation of the thermal regime to some depth in the permafrost is difficult to achieve using surface methods such as those described above. Highly efficient heat transfer devices known as thermo-piles or

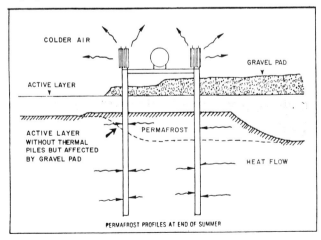

Figure 2. Control of thermal regime using thermo-piles. Adapted from Waters (131).

heat pipes have been used to cool the permafrost very effectively in winter. They function as a valve, however, and cease to extract heat in the warmer part of the year when the ambient temperatures exceed the ground temperatures. Waters (131) describes the devices used for support of a large percentage of the trans-Alaska oil pipeline (Fig. 2). During the summer period of inactivity, the ground around the piles can warm up almost to the adjacent undisturbed ground temperatures. Consequently, the designer should be cautious not to expect long-term reductions in the ground temperatures. The main benefit of thermopiles can best be viewed as preventing long-term thermal degradation of the permafrost initiated by construction disturbance, rather than lowering ground temperatures significantly in the long term.

THE FREEZE-THAW CYCLE - A GEOTECHNICAL PERTURBATION

The Effects of Frost Heave

Frost heave is the consequence of a complex coupling of heat and mass flow in a freezing soil system. Frost heave can be considered as having two components occurring together. The first is the nine-percent phase expansion of water contained in the pores of soil, depending on unfrozen water content effects (2). The second, and generally far greater component of heave, is the growth of ice lenses in the freezing zone as water is drawn to the freezing front from the still unfrozen soil.

When saturated coarser-grained soils with little or no fines are frozen in an open system, a volume of water associated with the phase expansion is usually expelled from the freezing zone. If the same soil is frozen in a closed system, with expulsion of water impeded, heave can result. Generally speaking, frost heaving due to this component is not a major design concern but, under certain conditions, frost heave and thaw settlement can occur even in coarse-grained soils.

The second and generally the most significant component of frost heave results when water is drawn to the freezing zone. Its severity depends upon a variety of conditions such as soil type, depth of water table, heat extraction rate and applied load. This drawing-up of water results from the pore water tensions created in a freezing soil system, and causes the overall water content of a soil to increase if a soil is frozen in an open system. Locally, however, the freezing process can tend to accumulate water as ice in the form of lenses, surrounding partially desiccated soil lumps which contain little or no ice and substantial unfrozen water content. The existence of unfrozen soil water requires, for thermodynamic compatibility, that the water be under a high tension. These tensions result in substantial increases in effective stresses within the soil as it continues to consolidate as water is drawn out of the soil to adjacent ice lenses. The lenses attempt to grow long after the freezing front has passed.

Thaw Settlement

In geotechnical practice, the term thaw settlement embraces the volumetric changes due to phase transformation and the subsequent con-

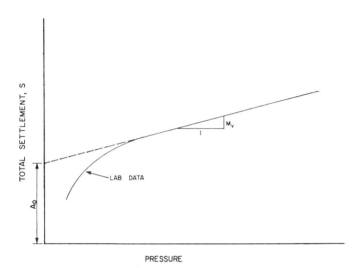

Figure 3. Generalized form of thaw settlement test data.

solidation of the soil as imposed stresses are transferred to the soil skeleton.

The total thaw settlement, S, associated with the one-dimensional thawing of frozen soil can be expressed in terms of

$$S = A_0 X + m_v \int_0^X (P + \gamma' X) \, dX \qquad (4)$$

where S is the total settlement, A_0 thaw settlement parameter, X depth to thaw front from original surface, m_v average coefficient of compressibility, P surcharge load, and γ' submerged unit weight of thawed soil. (In this equation it is assumed that the water table remains at the surface of the soil.) From a test on a given sample, settlement expressed as a vertical strain can be plotted against effective vertical pressure (Fig. 3). A best-fit linear relationship extrapolated to zero pressure defines A_0, and the slope of the line, m_v. This technique first appeared in the Russian literature (119) and has subsequently been used in North America by Watson et al. (123), Luscher and Afifi (1973) and McRoberts et al. (76). An alternative method of quantifying thaw settlement is to define the total thaw strain ε_p observed in a laboratory test at some effective stress considered representative of the in-situ stress that would exist after thaw settlement is completed.

Thaw Consolidation

When thawing occurs at a slow rate, the water generated will flow from the soil at about the same rate it is produced. This squeezing-

out is induced by the self weight of the soil or by applied load. If
the rate of water generation exceeds the discharge capacity, excess
pore water pressures will develop. This process and the study of the
relative influence of the rate of thaw, compared with the rate of consolidation, is generally referred to as thaw consolidation.

Important studies and theoretical developments regarding thaw consolidation were presented by Yao and Broms (134), Tsytovich et al.
(120), Zaretskii (135), and Lachenbruch (54).

These works contained various shortcomings, and it remained for
Morgenstern and Nixon (81) to set out the mechanics of the consolidation of thawing soil by coupling well-known theories of heat conduction
and of consolidation of a compressible soil. When frozen soil is subjected to temperature increase, thawing will proceed in a manner controlled by the temperature boundary condition and the thermal properties of the frozen and unfrozen soil. The solution to the heat conduction problem defines a region of thawed soil and the rate of movement
of the freeze-thaw interface. Classical Terzaghi consolidation theory
is then assumed to govern the rate of pore pressure dissipation within
the thawed zone. The freeze-thaw interface forms the lower boundary of
interest which moves with time, and hence the consolidation of the
thawed soil above this interface is governed by a moving boundary condition. At this thaw plane, water liberated by phase change flows upwards if excess pore pressure exists. For a saturated soil, the boundary condition requires that any flow from the thaw plane be coupled to
a change in volume of the now unfrozen soil in accordance with consolidation theory. The derivation of the boundary condition is given in
Morgenstern and Nixon (81). For a one-dimensional case in which the
thaw front was assumed to move according to equation 1, a solution for
the excess pore water pressures was obtained in terms of a thaw consolidation ratio R, or

$$R = \frac{\alpha}{2\sqrt{c_v}} \qquad (5)$$

where c_v is the coefficient of consolidation and α is as previously
defined. The parameter R expresses the ratio between the rate at which
water is liberated by the thaw front and the rate at which water can be
squeezed out of the soil skeleton. If this parameter approaches zero,
then no excess pore pressures are predicted. As R becomes high, the
pore water carries the entire overburden soil and the effective
stresses approach zero.

Several extensions to the original theory have been reported and
an excellent review is provided by Nixon and Ladanyi (91).

The initial effective stress in a soil thawed under undrained conditions has been termed the "residual stress" by Morgenstern and Nixon
(81) and as the "extent of initial consolidation" by Palmer (97). If a
soil is ice rich or has a high void ratio in the thawed, undrained
state, it is reasonable to assume that the residual stress is zero.
However, depending on the stress, thermal and geological history of a
frozen soil, significant residual stresses may exist. The higher the

residual stress, the smaller the thaw settlement, the lower the pore
pressures generated during thaw and the higher the undrained strength.
A detailed investigation of the role of residual stresses is given by
Nixon and Morgenstern (93).

The legacy of the freezing stage on the thaw component of the
cycle can now be considered. The manner in which excess water in the
form of ice is distributed in the soil will influence the magnitude of
the residual stress. If excess water is more or less uniformly distributed throughout the soil voids, then the residual stress is low.
Alternatively, the same volumetric expansion by heave in a different
soil, resulting in ice mostly in the form of lenses or veins separated
by denser soil nuggets, may yield a higher residual stress on thaw.

Thaw Weakening and Recovery

The effect of a freeze-thaw cycle on the strength and deformation
properties of a soil is determined by a complex interplay of freezing
and thawing rates, soil properties, drainage conditions and applied
loads. The interaction of these factors determines the effective
stress in the soil upon thawing and thereafter. In frost areas neither
the strength nor the resistance to deformation is a constant and
inherent property of a soil but both vary cyclically over an extremely
wide range. The dynamics of seasonal changes can produce frozen soil
with properties resembling soft rock, changing suddenly during the
spring thaw to a condition typified by the oft-applied expression "mud-
season," and then firming gradually as the soil reconsolidates and
desaturates through summer and fall with gradually increasing moisture
tension.

In general, moderate subfreezing surface temperatures, causing a
low rate of propagation of the frost line, create the conditions for
maximum growth of ice lenses fed by upward migration of moisture from a
proximate ground water table under the influence of pore water tensions
generated at the freezing front. In near-surface, fine-grained soil
with high pore water mobility, such as silt, these conditions lead to
an ice-rich frozen soil evidencing a large increase in volume, with ice
distributed throughout the soil as well as segregated in seams, veins
and lenses. Subsequent rapid thawing produces a saturated, highly
underconsolidated, and unstable soil having negligible strength and
residual stress, as well as a modulus of deformation (stress divided by
strain) near zero.

Variation in the conditions toward lower surface temperatures,
faster downward propagation of the freezing front, a more plastic soil
with reduced pore water mobility, and greater depth to groundwater
leads to localized moisture migrations from the surrounding soil to the
expanding ice crystals. The high pore water tensile stress in the clay
near the growing ice crystals partially desiccates the soil and creates
nodules of soil having increased effective stresses. The partially
desiccated nodules become overconsolidated, while most of the ice tends
to be concentrated in intervening shrinkage cracks between the nodules
(21). Most of the soil mass may actually exist at the moment of thawing in an overconsolidated condition, and the residual effective stress
will be reflected in significant strength and resistance to deformation.

A third set of conditions may be considered, with severe freezing
temperatures and water readily available from a high groundwater table.
The soil is a uniform medium to fine sand with a small percentage of
gravel. In this case, freezing comprises a phase change of the moisture existing within the soil pores. The expansion associated with the
phase change may cause either a slight heave, or expulsion of water
from the freezing zone. Subsequently, as thawing proceeds under moderate surface temperatures, readjustment of the soil grains and movement
of water within the thawed zones keep pace with advance of the thawing
front, with no excess pore pressure generated and the soil fully consolidated under the prevailing surcharge. In this case, the strength
and modulus of deformation in the thawed state are virtually unchanged
from the condition prior to freezing.

In the first two cases described here, both the shear strength and
modulus of deformation are substantially lower upon thawing than in the
pre-freezing condition. With the passage of the thawing front, the
thawed soil immediately begins a recovery process that will lead to
reestablishment of a pre-freezing condition, which, as noted above, may
represent successively greater degrees of overconsolidation, lower void
ratios, and consequently higher strengths. The recovery process initially is a process of reconsolidation. For soil existing above the
groundwater table, a desaturation process also takes place, resulting
in the gradual buildup of pore water tension. In near-surface, partially saturated soil, moisture tension (equivalent to soil suction) is
the dominant element in the intergranular effective stress, and consequently is the primary source of shear strength, resistance to deformation, and bearing capacity.

Freeze-Thaw Overconsolidation

The phenomenon of overconsolidation by freezing and thawing has
been reported by many workers. For example, Tsytovich et al. (120),
Malyshev (68), Nixon and Morgenstern (93), Chamberlain and Blouin (20)
and Chamberlain (18) all report significant increases in the consolidation of clay soils after one or more freeze-thaw cycles. This overconsolidation is caused by the tensions or negative pore water pressures
generated during the freezing cycle. These tensions create increased
effective stresses and consolidation of the soil adjacent to ice lenses
and below the freezing zone. When the soil is thawed in an undrained
state, a residual stress can be measured.

Unlike many other impacts of the freeze-thaw cycle which have negative effects, freeze-thaw densification may have positive design implications. For example, in a study of fine-grained dredge spoils,
which retain large volumes of water and consequently require large
areas for containment, Chamberlain and Blouin (20) showed that freeze-
thaw cycling would be an effective method of dewatering such material.

An example of freeze-thaw consolidation leading to increases in
in-situ shear strength can be found in vane data (Fig. 4) taken to
depths of up to 100 ft (30 m) within and along the margins of a shallow
lake in the discontinuous permafrost zone. At locations within the
present lake and with deeper water, the in-situ shear strengths in the

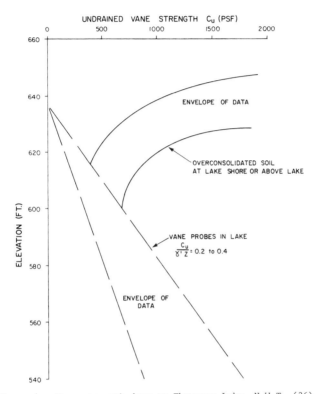

Figure 4.   Vane strength data at Thompson Lake, N.W.T. (36).

varved sediment tended to be linear with depth.   The undrained
strength/effective stress ratio is 0.2 to 0.4, with local variations
within the varved deposits.   The magnitude of this ratio and the fact
that the ratio is constant even close to the surface are characteristic
of a normally consolidated deposit.   However, vane data near the shore
and above the lake indicated substantially higher shear strengths within approximately 65 ft (20 m) of the surface.   It was concluded that
overconsolidation has occurred, attributed to the freeze-thaw cycling
of one or more episodes of permafrost.

MECHANICAL PROPERTIES OF THAWED SOIL

Bulk Densities/Water Content Relationships

        The frozen bulk density γ_f and the water content w of a thawed
soil are useful index properties.   If it is assumed that all water
present in frozen soil exists as ice, then the relationship is

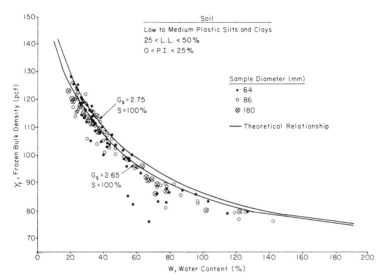

Figure 5. Correlation of frozen bulk density and water content (76).

$$\gamma_f = \frac{G_s \gamma_w (1 + w/100)}{1 + (1.09\ wG_s/100S)} \qquad (6)$$

where w is expressed in percent, G_s is the specific gravity of soil grains, γ_w is the unit weight of water and S is the degree of saturation expressed as a fraction. Several factors tend to reduce the γ_f observed for a given w. Figure 5 summarizes data observed in silts and clays of low to medium plasticity. Similar correlations are reported by Roggensack (107).

Thaw Settlement Parameters

In most applications, the designer will be required to undertake tests to establish thaw settlement parameters, because no predictive capability in terms of soil type, ice type and distribution, and bulk density has been developed. Nevertheless, extensive thaw settlement data accumulated for different projects enable the designer to obtain reasonable estimates of thaw strain without resorting to a major, expensive testing program. Data for soils along the trans-Alaska pipe-line route have been correlated with frozen dry density by Luscher and Afifi (61). Thaw strain was correlated with frozen bulk density in fine-grained lacustrine soils with a wide range of ice contents, at different points along the Mackenzie Valley, by Speer et al. (114). Tests reported by McRoberts et al. (76) include data on samples of 2.5, 3.4, and 7.1 in. (64, 86, and 180 mm) which gave essentially identical results. The results are plotted against total water content in Figure 6.

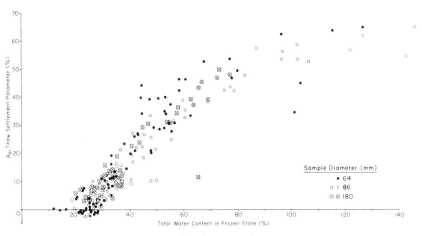

Figure 6. A_0 related to total water content.

Permeability and Coefficient of Consolidation

The magnitude of the coefficient of consolidation, c_v, can be determined directly by conventional interpretation of consolidation tests or indirectly by independent measurement of the coefficients of permeability, k, and of compressibility, m_v, and application of the well-known equation

$$c_v = \frac{k}{m_v \gamma_w} \; . \tag{7}$$

An example of this method was presented by McRoberts et al. (75) for a silty clay from two sites. Permeability data were obtained from laboratory tests on thawed samples and from in-situ permeability tests in thaw bulbs. c_v was calculated from relationships representing the average, lower bound, and upper bound of the permeability data, and estimates of m_v (Fig. 7). Comparison with direct measurements based on observing transients in accordance with Terzaghi consolidation theory shows reasonable agreement.

Discontinuities produced by melting ice lenses exert dominant influence on post-thaw fabric. The role of discontinuities and the influence on k and c_v are well established in geotechnical practice. For example, Rowe (108) presents a detailed study of the effect of fabric. Data on the consolidation of fissured clays indicate that secondary structure provides additional drainage capacity particularly at low stress levels (70). Roggensack (107) undertook direct permeability tests on thawed permafrost cores and found a nonlinear relationship between permeability and effective stress. At low stress levels, permeability was reduced substantially for small increases in stress level and this was attributed to the reclosing of cracks created

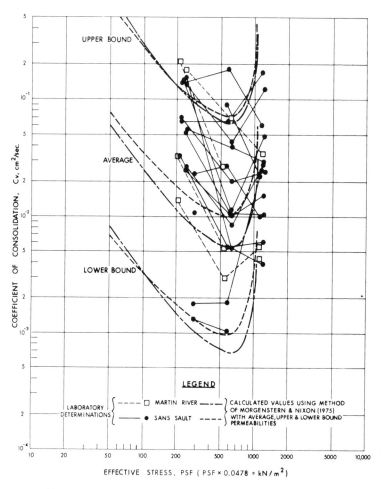

Figure 7.   Cv versus effective stress for Sans Sault and Martin River.

by ice lens formation. At higher stress levels, smaller decreases in
permeability with stress were found, attributed to conventional porosity decrease. Further data are reported by Chamberlain and Gow (21)
who clearly show the combined influence of the freeze-thaw cycle.
While the freezing cycle densifies the soil, the imparted structure,
which remains after thaw, increases the permeability. In some cases,
increases in permeability of up to 2-1/2 orders of magnitude were
observed.

Residual Stress

       As reviewed by Nixon and Morgenstern (94), two methods can be used
to measure the residual stress, σ'_o. In the first and direct method
the sample is thawed without drainage and the total stress continuously
adjusted so that the pore water stress is always zero. In this way,
the residual stress on completion of thaw is equal to the total stress
on the sample. The indirect approach (94) uses the relationship of
void ratio e vs log effective stress σ' from a thaw settlement test.
The sample is thawed under one stress increment and subsequent load
increments define the conventional curve. Based on the initial frozen
bulk density the thawed but undrained initial void ratio is calculated
and the e vs log σ' relation is extrapolated backwards to obtain the
effective stress at the initial void ratio. This stress can be considered to be the residual stress.

       Residual stress data are reviewed by Nixon and Ladanyi (91). Data
of Roggensack (107), summarized by Morgenstern (80), show some correlation between residual stress and liquidity index for clay and silt
soils. Figure 8 presents a summary of data on a sandy silty soil from

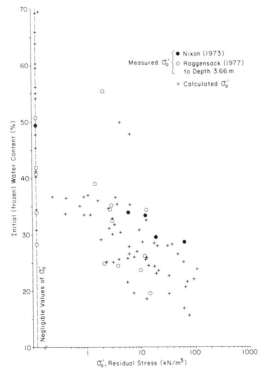

Figure 8. Comparison of residual stress determinations by different methods in the same soil.

Norman Wells. Shown in this figure are residual stresses reported by
Nixon and Morgenstern (93) and Roggensack (107) using the direct null
method, as well as the results of indirect calculations using data from
thaw settlement tests reported by McRoberts et al. (76). There is evi-
dently considerable scatter, much of which is due to minor variations
in soil type.

Excess Pore Pressures

An important geotechnical product of the thaw consolidation theory
is the prediction of excess pore pressures in the thawed zone depending
on the magnitude of c_v. Experimental proof of the theoretical predictions based on laboratory tests have been reported by Morgenstern
and Smith (83), Nixon and Morgenstern (94) and Roggensack (107). Field
evidence for excess pore pressures is reported for several sites in the
Mackenzie River Valley. Morgenstern and Nixon (82) report on excess
pore pressures measured beneath a buried warm oil pipe test section
operated at Inuvik, N.W.T. Piezometers were installed in ice-rich
clayey silt about 2 m below the warm pipe, and reasonable comparisons
of predicted and observed pore pressures are reported. Excess pore
pressures have been measured in naturally occurring thaw bulbs by
McRoberts and Morgenstern (73) and at different sites by McRoberts et
al. (75).

Shear Strength

For thawed soil, conditions prevailing during prior freeze-thaw
cycles determine the values of the residual stress and pore pressure,
and control the development of significant changes in the structure of
the soil, thus exerting a governing influence on the shear strength
during thawing and subsequently as recovery proceeds.

The first systematic laboratory investigation of the shear
strength of thawing soils was that reported by Broms and Yao (13).
This investigation included variation of conditions during prior freezing of a compacted silty clay, such as rate of freezing, surcharge and
ingress of water. Not surprisingly, high pore pressures were measured
during both unconsolidated-undrained and consolidated-undrained testing
of the thawed soil. It was found that pre-consolidation of the soil
during freezing was a principal factor influencing the shear strength
in the thawed state. In a following paper, Yao and Broms (134) reported that the reduction in strength is caused by an increase in water
content during freezing, a decrease in soil density and incomplete
dissipation of excess pore pressures during thawing.

An important development during the past decade of great signifi-
cance to an understanding of the shear strength of thawing soils is the
comprehensive one-dimensional theory of thaw consolidation worked out
by Morgenstern and Nixon (81). A key element of the relationships
postulated at the thaw line is the residual stress. Nixon and Morgenstern (93) showed the relationship of the residual stress to shear
strength. If drainage upon thawing is negligible, either because thawing occurs rapidly or drainage is impeded or occurs very slowly, then
the residual stress controls the initial undrained strength. For a
thawed cohesionless soil

$$\frac{C_u}{\sigma'_o} = \frac{(K_o + A\ (1 - K_o))\sin\phi'}{1 + (2A - 1)\ \sin\phi'} \qquad (8)$$

where C_u = the undrained shear strength of thawed soil

σ'_o = the residual stress

K_o = the ratio between the lateral and vertical effective
stresses under conditions of no lateral yield

A = the pore pressure parameter

σ' = the angle of internal friction, based on effective stress.

But some drainage and partial consolidation usually occur during
thawing. Consequently the shear strength depends not only upon the
residual stress but also the thaw consolidation ratio R. The process
of recovery from the thaw-weakened state consists not only of reconsolidation and dissipation of positive excess pore pressure but, in the
case of soil lying above the groundwater table, also includes a desaturation phase with development of pore water tension, as described
below. As moisture tension builds up during desaturation, the recovery
of shear strength can be approximated in the context of established
relationships for partially saturated soils (32).

Few laboratory measurements of the shear strength of thawed soil,
and still fewer cases of validation or comparison of laboratory results
with field performance, have been reported. Culley (26) reported an
intensive series of tests on a sandy silty clay (glacial till) compacted at various moisture contents, and subjected to repeated-load
triaxial tests before and after closed-system freeze/thaw. It was
found that both recoverable and non-recoverable strains were higher and
volume change lower for specimens compacted at lower moisture contents.

The undrained strength of thawed soils was measured during the
1970's by several investigators. Watson et al. (132) obtained samples
of an ice-rich silt permafrost and investigated its undrained strength
after thawing. Samples were thawed and consolidated under pressures
from 400 to 1800 psf (19 to 86 kPa), and the tested in the vane shear,
unconfined compression and consolidated undrained modes (Fig. 9).

Roggensack (107) conducted triaxial tests on thawed specimens of
two permafrost soils, a silty clay and a soil consisting of interbedded
silt and clay. The undrained strength was found to be directly
related to the residual stress.

Nixon and Hanna (90) reported unconfined compression tests on
thawed specimens of natural permafrost soils including a non-plastic
silt and a silty clay glacial till. The overall range of shear
strengths, taken as half the unconfined compressive strength, was 0 to
6.5 psi (0 to 45 kPa). A rough correlation was found with frozen bulk
density (Fig. 10), and with depth. The results were consistent with
measurements in earlier investigations indicating that at frozen bulk
densities less than about 106 pcf (1700 kg/m^3) the residual stress
tended to zero. The results also suggested that if the thaw strain is
less than about 10%, some undrained strength might be anticipated.

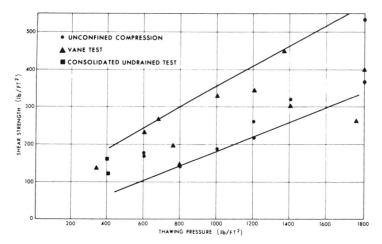

Figure 9.   Undrained shear strength of thawed and consolidated specimens of ice-rich silt permafrost (132).

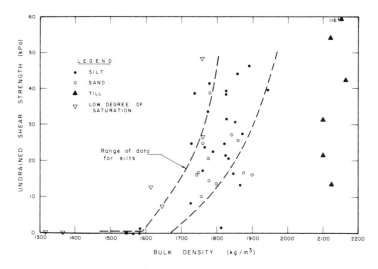

Figure 10.   Unconfined compression test data for thawed soils correlated with frozen bulk density (90).

Alkire (1) compared the effects of cyclic loading and freeze-thaw cycling on the unconsolidated undrained strength and the strain at failure of a silt of low plasticity, concluding that stress cycling and freeze-thaw cycling produced similar response in terms of pore pressure development and strength. The strain at failure was much greater for samples subjected first to freeze-thaw cycling. Udd and Yap (124) investigated the strength of thawed specimens of naturally frozen iron ores. Both uniaxial compression tests and direct shear tests showed severe strength reductions upon thawing.

The results on thawed samples mentioned above apply to axial compression tests performed on specimens after thawing was complete and after equilibrium was reestablished to varying degrees. Shear strengths during thawing, and in particular those at the thawing interface, which doubtless are lower, have been little investigated. Thomson and Lobacz (118) developed a technique for direct shear tests with control of the thaw interface to coincide with the shear plane. Sage and D'Andrea (109) developed a simple shear device for testing thawed specimens and in a continuing investigation have tested various soils before and after freeze-thaw.

Resilient Modulus

The concept of the modulus of resilient deformation was established by Seed et al. (111) as the applied stress divided by the recoverable strain. It is a useful concept for the design of roads, airfields and railroads, in which repeated application of loads at the surface produces stress in the underlying soils well below the limit state for monotonic loading, causing repeated recoverable strains throughout the layered system and potential fatigue failure in certain pavement layers.

The resilient modulus, like the shear strength, is very high for soils in the frozen state, dropping precipitously upon thawing and recovering gradually through late spring, summer and fall as the soil first reconsolidates and then desaturates (Fig. 11). Like shear

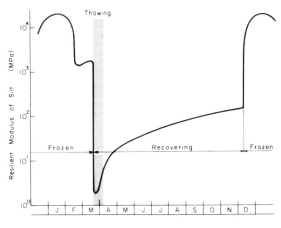

Figure 11. Seasonal variation in resilient modulus of a silt (45).

strength, the resilient modulus during thawing is lowest in soils that
had been ice-rich in the frozen state. This condition is governed in
part by the properties of the soil, and in part by conditions prevailing during freezing.

Laboratory repeated-load triaxial stess on various fine-grained
soils (clays, sandy clays, and silty clays) after laboratory closed-
system freeze-thaw have been reported by Pagen and Khosla (95), Culley
(26), Mickleborough (77), MacLeod (66), Bergan (10), Bergan and
Monismith (9), and Robnett and Thompson (106). The reported resilient
moduli range from about 1500 to 7500 psi (10 to 50 MPa). The dominant
influence of prior freezing conditions upon these moduli is evidenced
by the results of laboratory repeated-load triaxial tests on specimens
cored from frozen road subgrades in which freezing had occurred without impediment to ingress of water (46, 23); these tests gave resilient
moduli an order of magnitude lower, ranging from 150 to 600 psi (1 to 4
MPa). Similar results were reported for resilient moduli calculated
from in-situ plate bearing tests.

Resilient moduli for cohesive soils are somewhat dependent upon
the magnitude of the deviator stress, and may be expressed as:

$$M_r = K_2 + (K_1 - \sigma_d) \ K_3, \ \sigma_d < K_1 \qquad (9a)$$

$$M_r = K_2 + (\sigma_d - K_1) \ K_4, \ \sigma_d > K_1 \qquad (9b)$$

where M_r = resilient modulus

σ_d, the deviator stress in triaxial tests = $\sigma_1 - \sigma_3$

K_1, K_2, K_3, and K_4 = constants.

Nevertheless, the variation of the resilient modulus with deviator
stress is relatively slight, and is sometimes neglected, considering
the behavior of such cohesive soils to be linear (independent of
stress). Cohesionless soils are more strongly nonlinear, the resilient
modulus being dependent upon the stress invariants according to relationships of the form

$$M_r = K_1 \ J_1^{K_2} \qquad (10)$$

or (24)

$$M_r = K_1 \ (J_2/\tau_{oct})^{K_2} \qquad (11)$$

where J_1, the first stress invariant = $\sigma_1 + \sigma_2 + \sigma_3$

J_2, the second stress invariant = $3\sigma_3^2 + 2\sigma_3\sigma_d$

τ_{oct}, the octahedral shear stress = $\dfrac{\sqrt{3}}{2} \ \sigma_d$

K_1, K_2 are constants.

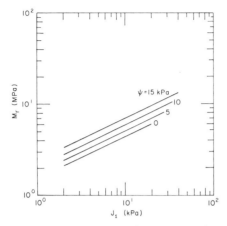

Figure 12. Dependence of resilient modulus M_r of silty sand upon first stress variand J_1 and soil moisture tension ψ (47).

If drainage of excess water is not impeded, the soil upon thawing begins a recovery process, consisting initially of reconsolidation, involving the dissipation of excess pore pressure. In the pavement design problems to which the concept of resilient modulus is applied, the soil of interest is above the groundwater table. In this case reconsolidation is followed by gradual desiccation, involving the buildup of pore water tension. Chamberlain and Cole (22) described a triaxial test technique for evaluating the resilient modulus throughout the recovery process. The technique has been applied to several silty sands (24, 47), for which the regression constants were evaluated in terms of the dry density of the soil and the soil moisture tension. The dependence of resilient modulus of one of these soils on moisture tension and a stress parameter is shown in Figure 12.

Moisture Tension

Near-surface soils commonly exist in a partially saturated condition in which the soil moisture is in a state of tension, under the influence of evaporation which causes the groundwater table to be drawn below the ground surface. Under these conditions the total stress at a given depth will have remained essentially constant while the pore water pressure became negative, giving rise to consolidation of the soil and to progressive desaturation. Fredlund (30) pointed out that various plants, by drying the soil in a process of evapotranspiration, are capable of applying 10 to 20 atm (1 to 2 MPa) of tension to the soil moisture before reaching their wilting point. Moisture tensions of less than 1 atm (100 kPa) can be measured at various moisture contents in the laboratory by means of Tempe cells (37). A variety of devices have been used to monitor the variation of moisture tension in field conditions, not all of which resist damage by freezing. McKim et

al. (71) developed a tensiometer that has been used successfully in a
freezing environment to measure moisture tensions of less than 1 atm
(100 kPa).

By increasing the effective stress, the tension in the soil moisture strongly influences the shear strength and resilient modulus. In
near-surface soils the effective stress is approximately equal to the
soil moisture tension, which is large in relation to the overburden
pressure (14).

Moisture tension is not constant throughout the year but varies in
relation to changes in temperature and precipitation. In frost areas
two periods of increasing moisture tension can be identified, and in
both cases the increasing moisture tension has a favorable effect on
soil strength and deformation properties. The buildup of moisture
tension during freezing of moist soil, the resulting increase in effective stresses and overconsolidation of the soil, and the favorable
effects on residual stresses upon subsequent thawing, were noted above.
With passage of the thaw front the moisture tension at a given depth
commonly is reduced to zero, and positive pore pressures may be noted.
After gradual dissipation of pore pressures, moisture tension builds up
through late spring and summer, interrupted on occasion by periods of
precipitation, but generally reaching a maximum at the onset of another
freezing cycle.

GEOTHERMAL DESIGN FOR FACILITIES SUBJECTED TO FREEZING AND THAWING

Preserving the Frozen Condition

The design approach involving preservation of the thermal state of
the permafrost, sometimes known as the passive method, is adopted where
thawing would lead to unacceptable settlements. The frozen state is
maintained by placement of an insulated pad for a non-heated structure,
or by ventilation and insulation for a heated structure. Ventilation
can be achieved by elevating the structure and maintaining a 3- to 4-ft
(0.9- to 1.2-m) air space between the heated building and the ground
surface. According to Lobacz and Quinn (59), a cooling trend takes
place if the airspace is constructed correctly and maintained clear of
obstructions in winter.

Where heavy floor loads will be experienced in warehouses, ga-
rages, hangars or heated tanks, ventilation is achieved by introducing
a system of ducts in an insulated pad (4, 87). Air flows either naturally through open-ended ducts, or is forced through manifolds which
supply air to the buried ducts. It can be shown that natural ventilation induced by the prevailing wind velocities is usually inadequate
for larger on-grade heated structures, which usually require forced
ventilation using fans. The geothermal design of such foundations is
treated by Nixon (87), and involves two primary aspects, namely the
provision of sufficient granular fill and insulation to prevent thaw
into the permafrost in summer, and secondly establishment of adequate
heat removal in the form of ventilation to refreeze and chill the pad
in winter. Typically, for pads containing 4 in. (10 cm) of synthetic
insulation, air flow rates must be in the range of 0.03 ft^3/s per ft^2

(0.01 m^3/sec per m^2) of floor area to prevent excessive warming of the air near the outlet end of the ducts.

Insulated pads may be used in some cases to preserve the frozen condition of the permafrost, but in sensitive, discontinuous permafrost where the mean ground temperature is -3°C or warmer, no reasonable thickness of gravel and/or insulation may be sufficient to completely prevent some slow thawing of the permafrost. In these cases some active form of heat removal beneath the pad surface is required to achieve thermal stability. The development of "thermo-piles" has provided a very efficient system of heat removal from the ground, with no moving mechanical parts and little maintenance requirements (Fig. 2). The best-known application of this technology was in the construction of the trans-Alaska oil pipeline, where thermo-piles were used for the aboveground pipe supports to prevent long-term thawing.

Permafrost preservation for a buried hot oil pipeline or vertical oil well can normally be achieved only by combined insulation and refrigeration. This is usually very costly and, alternatively, elevated pipelines with thermo-piles, forced air ventilation, or shading might be used to maintain the permafrost thermal regime.

Pre-thawing of Frozen Ground

Pre-thawing of the ground prior to construction has been referred to as the active method of foundation treatment (48). In general, it has not received wide acceptance, due mostly to the great expense in providing the large amounts of heat necessary to thaw the soil. Typical latent heats for a permafrost soil with 30% water content by dry weight are in the range of 4000 Btu/ft^3 (150 x 10^6 J/m^3), and this, in addition to the sensible heat of the soil, must be supplied in a timely fashion to the volume of soil under consideration. Depending on the coefficient of consolidation of the soil, and the method of thawing, a large percentage of the settlement may occur during thawing, but a certain "post-thaw" period of soil consolidation and stabilization must be planned. For example, a layer of silty clay 15 ft (4.5 m), thick, and having a consolidation coefficient, c_v, of 0.01 cm^2/s will require in the range of 8 months for complete consolidation.

Pre-thawing can be carried out by surface electrical heating, or by the use of steam points. Rapid thawing of this nature in fine-grained soils will result in soft, weak, underconsolidated soils. These methods are expensive, and are only suitable for upgrading of relatively small volumes of frozen subsoils for building foundations. Electrical methods may also prove useful in certain circumstances where it is necessary to maintain subsoils in an unfrozen condition beneath unheated or cold structures.

The thermal effects of flowing water can also be used to advantage in pre-thawing. It was proposed that the foundation (in a frozen, jointed rock) of a large dam in the U.S.S.R. be thawed by using pumped river water at 39-41°F (4-5°C).

If time is available, stripping and clearing of the surface vegetation and surface organic layers may be used to deepen the permafrost

table in a discontinuous permafrost area, and to pre-thaw the near-surface layers. These are often the icier layers, and this method could
be successful in eliminating a large percentage of the total settlement. To evaluate the suitability of the procedure for a given subsoil
condition, the data of Linell (56) can be consulted. The depth of thaw
for the continuously cleared site can be approximately related to the
square root of time by equation (1), evaluating the thaw parameter α as
approximately 4.3 ft/yr$^{1/2}$ (1.3 m/yr$^{1/2}$) for fine-grained soils.
In coarser-grained soils, α might be larger due to a lower latent heat
term and higher thawed conductivity. According to this relationship,
thaw depths of 9 ft (2.7 m) might occur in a 4-year period, indicating
that this approach clearly requires a long lead-time to achieve rather
nominal thaw depths. But the advantage is that the pre-thawing is
achieved at minimal cost and effort, and consolidation will likely
proceed concurrently with thawing. Esch (29) reported on the use of
thin gravel pads, surface polyethylene sheeting and dark painted surfaces to deepen the permafrost table prior to highway construction in a
discontinuous permafrost area. The depth of thaw has increased to
about 7 ft (2 m) in the 2 years that the test sections have been under
observation.

Preservation of Thawed Conditions

If an area has been thawed, or was unfrozen in the first instance,
the thawed condition beneath unheated structures or roads can be maintained by using relatively conventional measures for seasonal frost
protection. These include economic combinations of granular fill and
synthetic insulation. The same procedures that are detailed below for
design of work pads in permafrost terrain can be applied almost directly to the design of insulated pads for seasonal frost protection.

Freeze-thaw Cycling

If any significant depth of subsoil other than clean granular soil
beneath a pavement, pad or structure is allowed to freeze and thaw on a
seasonal basis, some seasonal motion of the structure is to be expected. Frost heaving and subsequent softening or weakening during thaw
are the consequences of seasonal freezing of finer-grained soils and
must be considered in geotechnical designs.

If seasonal frost advances below the base of a foundation, then
basal heaving of the foundation may result even under rather heavily
loaded foundations. Another potentially serious effect is the frost
jacking or gripping of vertically aligned structures such as posts,
piers, and piling. When designing piling in seasonal frost or permafrost zones, it is permissible to allow seasonal freezing and thawing
around the upper reaches of the pile, provided embedment in permanently
frozen ground or permanently thawed ground is sufficient. A common
rule-of-thumb for permafrost areas, which appears to have been based on
Russian literature, is that the minimum depth of pile embedment should
be three times the depth of seasonal frost. This is based on the fact
that the strength of the underlying permafrost is typically about one-
half of the tangential adfreeze bond exerted upwards by the rapidly
freezing active layer. In areas of softer unfrozen subsoils, or in
warm discontinuous permafrost, the embedment depth based on this simple

criterion might have to be increased significantly, depending on the ratio of the adfreeze bond to the strength of the underlying soil-pile contact.

In the case of unheated structures that may experience seasonal frost action beneath the footings or slabs of the foundation system, other possible alternatives include the design of extremely rigid structures that can withstand differential subsoil movements. Such foundations will usually become prohibitively expensive for anything other than relatively small structures. Lightly loaded temporary or expedient structures can incorporate leveling devices such as jacks or shims to accommodate differential vertical heave and settlement.

GEOTECHNICAL DESIGN FOR SUBSOIL THAWING

Foundations

The design of shallow foundations for thaw conditions proceeds in a manner analogous to conventional practice, and similarly it will be found that deformations rather than bearing capacity usually govern the design. Strengths for bearing capacity analysis may be assessed by using an effective stress approach, including an accounting of excess pore pressures, or alternatively by undertaking undrained strength tests on frozen samples thawed in a triaxial cell. If a foundation stratum with a high residual stress is available, it may be practical to place foundation systems directly on permafrost without concern for subsequent thaw. The designer should also note that, if the residual stress is higher than the combined applied load and soil weight, then thaw swell, rather that settlement, may occur (25). However, if much loss of strength on thaw is predicted, it will usually be found that settlements govern and may be excessive for the structure. Practice in the USSR in the design of foundations on thawing soils is presented in a recent symposium proceedings (126).

Pile foundations can be used to transmit loads through thawing subsoil if a competent bearing stratum can be located at depth. The bearing stratum can be unfrozen sediments, bedrock, or permafrost that will remain frozen over the structure's life as predicted by geothermal analysis. Piles used in this configuration must be designed for negative skin friction in the thawing zone. It must be remembered that the thawing zone will gradually increase over the design life. If bearing capacity is being achieved in the lower permafrost, attention must be directed towards creep effects (92). In many instances where piles are used to reach a competent soil stratum at depth, near-surface icy soils that are permitted to thaw may seriously complicate the design because of inadequate lateral support or even soil liquefaction.

Subsoil thawing may also become a design consideration for a variety of structures in which freezing has taken place either by design or by accident. Examples are the rehabilitation of old freezing storage structures for foodstuffs or liquid natural gas, or of recreational facilities such as hockey and curling rinks. Freezing is also used for temporary support for a range of applications such as tunnels, shafts, and retaining walls. In all these cases, as reviewed in detail by

Chamberlain (18), it is necessary to consider the nature of the frozen ground and the likely effects of thawing. The structures may have heaved during the freezing stage and become damaged. However, even more damage can result during the thaw cycle due to settlement and loss of bearing strength.

Work pads

In the past, many road projects in the Arctic have employed 3 to 6 ft (1 to 2 m) of fill, depending on permafrost conditions and local experience. Some thawing of the underlying permafrost inevitably takes place in all but the most northerly latitudes. Long-term problems due to frost action in the active layer, or to continual deepening of the active layer in discontinuous permafrost, necessitate costly annual maintenance programs.

Such problems have been alleviated in the past by the use of thicker embankments. Depending on the proximity of a borrow source, gravel availability, access, magnitude of the project, and other economic factors, the use of synthetic insulation may become clearly economical (85, 69).

The prime consideration in the design of insulated pads involves a thermal analysis of a layered soil system consisting of gravel fill, insulation, and different subsoil layers. The complex nature of the layered system and the variable surface thermal conditions usually require that these problems be solved by numerical methods on the digital computer. One important exception to this is an elegant analytical solution to the problem of a three-layer system subjected to a surface temperature sine wave presented by Lachenbruch (54). The principal restrictions to use of this solution in discontinuous permafrost are that there can be no phase change in any of the three layers and that the solution does not model the gradual change in unfrozen water content below 0°C in the permafrost subsoil. However, the solution has been used with reasonable success in cases where the gravel fill and insulation may be assumed to be completely dry and the 0°C isotherm remains above the permafrost.

Selecting reasonable thermal properties for gravel fill, polystyrene insulation, and frozen ice-rich silt, Lachenbruch's solution to the insulated fill problem was given by Nixon (88) and is shown on Figure 13. The results are presented in terms of the dimensionless ratio between the mean annual pad surface temperature (in degrees Celsius below freezing) and the amplitude of the sine wave, A_0. The assessment of the surface sine wave characteristics is complex and requires great care (125, 35). Once the temperature wave is known, however, Figure 13 can be used to estimate the required thicknesses of fill and insulation to maintain the 0°C isotherm at or above the pad base.

It may not be desirable or feasible to obtain complete thermal protection of the permafrost, and some limited amount of thaw each season may be deemed tolerable by the designer. The insulation and gravel thickness criteria for the pad may be established by carrying out a number of one-dimensional computer simulations using different

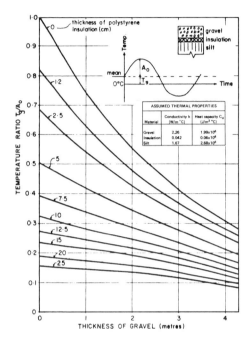

Figure 13. Lachenbruch analysis for thaw
or frost protection by means of insulation and gravel cover.

thicknesses of gravel and insulation. If some partial thawing is
deemed tolerable, the bearing capacity of the thawing subgrade layer
must be checked. Bearing capacity problems are usually avoided by the
provision of some minimum total pad thickness of the order of 3 to 4 ft
(1 to 1.2 m), depending on subgrade soil properties.

Experience with performance of insulated pad sections indicates
that it is thermally more efficient to keep the insulation as high as
possible within the pad section. However, there is a minimum depth of
granular cover that must be placed over an insulation layer to protect
it from being overstressed due to construction and service traffic
loadings. A partial stress analysis may be carried out by calculating
the stress distribution beneath the wheel load using the Boussinesq
theory, and the required depth of cover to limit the stress in the
insulation can be estimated as shown on Figure 14. More complete
analyses are outlined in the next section.

Murfitt et al. (85) provide a good review of design practices for
roads or pads on thawing permafrost or muskeg soils. Fill thicknesses
in excess of desired grade levels are recommended to account for subsidence caused by low strengths of thawing soils during summer con-

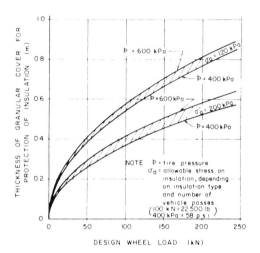

Figure 14. Granular cover requirements for
different wheel loads (88).

struction. The cone penetrometer is used to provide soil strength
index values for design. When placing synthetic insulation, backfilling using the "rolling surcharge" technique is recommended. In
summary, construction recommendations offered by Murfitt et al. (85)
included leaving the organic mat intact, avoiding cut sections in icy
permafrost, maintenance of micro-drainage systems, and stabilization of
cuts and fills to control thermal and hydraulic erosion.

Roads and Airfield Pavements

The objective of the design of pavements for subsoil thawing is to
dimension the layered system and choose materials appropriately to control the distress modes of cracking and distortion under wheel loads.
Without effective control, accelerated distress may occur because of a
decrease in strength and increase in deformability during thaw. At the
AASHO Road Test, the design of the experimental pavement was such that
distress was not effectively controlled, with the result (Fig. 15) that
damage under continuous application of truck axle loads occurred at
accelerated rates when the load applications coincided with spring
thaws. Approaches for design of pavements to resist other distress
modes are beyond the scope of this paper.

In both seasonal frost and permafrost areas, pavement distress is
controlled in part by limiting the selection of materials for the upper
layers to strong materials of low deformability that are not adversely
affected by the freeze-thaw cycle. For all but low-traffic-volume
roads, the uppermost layer is stabilized with a binder, usually
asphalt. Stabilized materials are susceptible to fatigue distress.
Repeated application of stress well below the limit state for monotonic
loading causes cumulative damage, leading eventually to cracking fail-

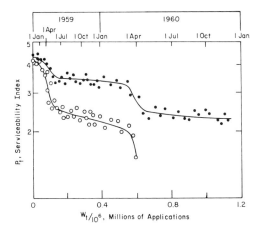

Figure 15. Deterioration of asphalt-paved
test roads with traffic and seasonal changes
(96).

ure in fatigue. Rutting, or permanent deformation manifested in the
surface course, is also a cumulative distress mode and, in pavements
with well designed surface and base courses, is usually considered to
originate mainly in permanent deformations in the subgrade.

In the past two decades it has become feasible to dimension layered systems from a mechanistic approach, that is, on the basis of calculated stresses, strains and deflections. It has also become feasible
to account for the cumulative nature of damage under successive load
applications in a more rational way. Earlier, through research and
observation of pavement performance, it had been learned that pavements
could be dimensioned appropriately to perform well under a given se-
quence of traffic loads in terms of index values of the soil's supporting capacity, such as CBR or R-values.

In the traditional design methods, still used extensively today,
cumulative damage has been recognized and treated by the dependence of
the pavement structure thickness on the number of load repetitions.
The seasonally varying rate of damage accumulation, caused principally
by changes in soil support during freeze-thaw cycling, is not explicitly treated. Instead the total accumulation over a period of several
years has been estimated, basing the design on a soil support value
selected to be applicable to average annual conditions. Or, in some
design methods the dimensioning is based on the critical spring thaw
condition, during which the rate of damage accumulation reaches a maximum. The traditional approach is exemplified by the design criteria
presented by Linell et al. (56).

Methods are available currently for design of pavements based upon
annual accumulation of damage that occurs at widely varying rates
depending on temperatures and freeze/thaw cycles. The approaches were

reviewed by Johnson et al. (44). The essence of the methods involves
the division of the year into intervals during which the temperature-
or freeze/thaw-dependent properties of each layer may be taken as
invariants. For the material properties prevailing during each interval, the horizontal strain at the bottom of the surface course and the
vertical strain at the top of the subgrade, caused by a single load
application, are calculated. Percentages of fatigue life of the surfacing layer and of the allowable rutting damage that are consumed by
the total number of load applications during each interval are estimated. And finally, the fatigue and rutting damage are summed for an
annual cycle, and the number of years projected to reach 100% of the
allowable damage of each type is compared with the desired economic
life of the pavement.

A study of a highway near Regina, Saskatchewan, by Bergan and
Monismith (9) is an example of application of seasonal variation in
material properties to the analysis of the cumulative effects of traffic leading to fatigue failure of a road in a seasonal frost area.

The calculation of the strains can be performed using any of
several available formulations of the layered elastic system problem.
Computer programs CHEV 5L (Chevron Research Company) and BISAR (Shell
Oil Company) are widely used. For pavements that include a nonlinear
layer or subbase an iterative approach is needed; the objective is to
determine resilient moduli and stresses for the nonlinear layer that
are compatible with the nonlinear characterization of the granular
material in the layer. A deficiency of the layered elastic system
approach is the inability of the system to accept horizontal variation
of the resilient modulus, even though the stresses are known to vary.
A partial solution of this problem was developed by L.H. Irwin in a
special formulation of the layered elastic system termed NELAPAV (24).
An approach using the finite element method also has been available for
some time (28). The finite element technique is less economical of
computer time but has the distinct advantage of accounting for radial
variation of moduli of nonlinear materials. Raad and Figueroa (105)
presented an improved finite element idealization of pavement systems
(ILLI-PAVE) that incorporates nonlinear behavior of granular soils and
a Mohr-Coulomb failure criterion. Further improvements were developed
by Brown and Pappin (15) in SENOL, a version that relies on a more
detailed method of characterizing granular materials and automatically
redistributes stresses when failure is approached in certain elements.

Calculation of fatigue damage requires either that fatigue tests
in flexure be performed on specimens of the asphalt concrete taken from
the surfacing layer or that fatigue data on asphalt concrete reported
in the literature be used to approximate a relationship between the
flexural strain, ε_h, under a single load application and the number
of repetitions to failure. A design procedure developed by Barker and
Brabston (6) makes use of such a relationship (Fig. 16). To estimate
the fatigue life consumed during each time interval, the number of
repetitions to failure is taken from the fatigue characterization; the
consumption of fatigue life is then the ratio of the actual repetitions
during the interval to the repetitions to failure.

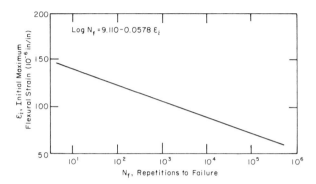

Figure 16. Fatigue life of flexural specimens of asphalt concrete (6).

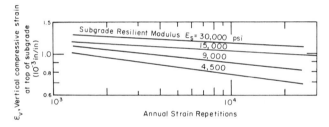

Figure 17. Subgrade strain criteria for critical aircraft traffic areas (6).

The analysis of permanent deformation makes use of the calculated vertical strain, ε_v, at the top of the subgrade. Again, the approach is illustrated (Fig. 17) by reference to the design procedure of Barker and Brabston (6), which justifies the adoption of subgrade strain criteria differing from those developed by other investigations. The percentage of the allowable rutting damage accumulated during each time interval is estimated in a manner similar to that described for the fatigue analysis.

After completing the first cycle of calculations the trial pavement section being analyzed is adjusted as necessary until the calculated times to failure in fatigue and in rutting become equal to the desired economic life of the pavement.

Several comments on the materials used in the layered system are offered, with particular reference to their function and performance during the thaw period. In the traditional approach to design of roads and airfields subject to freeze-thaw cycling, no frost-susceptible soils are used within the base and subbase courses. From the point of view of pavement performance during thawing, however, it seems clear that this is not an inviolable restriction, and that its application

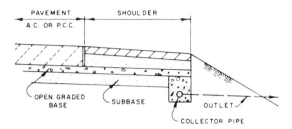

Figure 18.   Drainage system for frost areas (17).

should be judged on the basis of stress/strain analysis. In cases
where clean granular soils are scarce and costly, it may be economical
to utilize more readily available sands and gravels containing significant percentages of fines, limiting their use to lower layers where
lower resilient moduli during thaw periods have less impact on pavement
performance.

Traditional practice has admitted a moderate content of soil fines
in the upper base layers, as long as the material is not judged to be
frost susceptible. Yet Johnson (41) notes that even a small amount of
fines aggravates the adverse effect of saturation on the dynamic load
response of graded aggregate base. Lovering and Cedergren (60) showed
that only a small percentage of fines severely reduces the coefficient
of permeability of well-graded sand and gravel. Since the uppermost
base course will be the first layer of soil to become thawed in the
spring, its free-draining properties are essential to permit thaw consolidation, dissipation of pore pressure, and recovery of resilient
modulus in the layers below as they thaw. Hence it will usually be
economical to require, as a drainage layer, an open-graded upper base
course essentially free from fines,and possibly bound by up to 2 percent asphalt as advocated by Cedergen/KOA (17) (Fig. 18).

It is also pertinent to comment upon the use of insulating layers
to control freezing, and consequently also thaw-weakening, of pavement
subgrade soils. The approaches outlined herein for work pads are
generally applicable to pavements as well. For pavements, the cover
required over insulating layers to meet structural requirements may be
greater than that applicable to gravel work pads. Full-scale tests
reported by Barker and Parker (7) led to the conclusion that for pavements consideration must be given not only to the compressive strength
of the insulation material but also to the resilient strains in the
insulation and in the layers above and below it, ensuring that such
strains do not exceed allowable subgrade strain criteria (see above).

Mention should be made also of a developing technique for control
of both frost heave and thaw weakening in pavement subgrades, by encap-
sulating one or more layers within an impervious membrane. If fine-
grained soils are encapsulated at a moisture content several percent
drier than the optimum for compaction, frost effects are significantly
reduced (104, 113).

Pipelines

The design of buried pipelines where thawing of the permafrost occurs is a difficult and complex task, challenging the expertise of the arctic geotechnical engineer, the terrain analyst and the structural engineer. The following subsections briefly review the major design issues to be addressed by the designer of a pipeline that could cause thawing of the ground above, adjacent to, and beneath the pipe.

(a) Thaw Settlement

A geothermal analysis is needed to indicate the extent of the thawed zone around the pipe, and how it develops with time. The "design life" of the pipe often becomes an important issue to be settled as the basis of thermal calculation, as thawing usually progresses deeper with time. The analysis will reveal that some depth of permafrost, typically from 10 to 50 feet (3 to 16 m), will thaw over the design life of the structure. This fact aids in establishing the depth of subsurface exploration that must be achieved by drilling, geophysical surveying and interpretation of surficial geological features. On a larger project, samples are normally collected from the predominant stratigraphic units, and subjected to thaw strain testing to obtain the thaw settlement parameter, A_0, and the coefficient of compressibility, m_v. The settlement of a layer of soil can be estimated by knowing the effective overburden stress, σ', and the thaw settlement parameters (91). The settlement for an 'n'-layer profile can be calculated as

$$S = \sum_{i=1}^{n} (A_o + m_v \sigma_i') H_i \qquad (12)$$

where σ_i' is the effective overburden stress at the mid-point of layer "i", H_i is the height of layer i, and the summation commences at the first soil layer below the pipe base.

This summation can be carried out by assuming the pipe base is located at different depths below the ground surface to indicate (possible) beneficial effects of deeper pipe burial.

It is not feasible to predict the continuous settlement profile that the base of the pipe trench might experience, as the settlement can change significantly over short distances (114). It is practical and expedient to consider the dominant landforms within each physiographic region along the pipeline route, and to make a settlement estimate for each terrain unit, based on the boreholes within that unit.

If the settlements were uniform, then there would be little cause for concern for strains within the pipe. However, it is fairly obvious that if an area of unfrozen, or thaw-stable, ground or bedrock exists anywhere within the terrain group, then the differential settlement might easily equal the total settlement. In addition, as precise boundaries between thaw-settling and thaw-stable ground will not be known precisely, it must be assumed by the designer that the differential settlement might occur in a sudden "step" fashion. While this

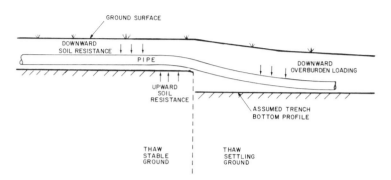

Figure 19.   Differential thaw settlement at a transition.

approach is conservative, without more complete documented case records
at thaw–settling transitions there is little justification for other
assumptions. The existence of stable soil above the permafrost, as at
some off–shore locations, would favorably affect the abruptness of the
differential settlement.

In addition to the magnitude of the differential settlement at the
transition, the geotechnical engineer must prepare soil load–deformation relationships and soil overburden loading conditions for use by
the pipe structural designer. These assessments can be made by considering the pipe as a strip footing, and calculating the subgrade reac-
tion in accordance with Terzaghi (117) or other works.

The pipe is normally considered as a Winkler beam on an elastic-
plastic foundation. Commercially available computer programs have been
extended to account for the loss of support within the thaw–settling
zone, the vertical overburden loading and other effects. Of critical
importance is the vertical "spring" in the stable soil immediately
adjacent to the thaw–settling zone (Fig. 19). If this "spring" is very
stiff, then the differential settlement that the pipe can tolerate may
be quite small, of the order of 6 to 18 in. (0.15 to 0.45 m). If this
spring is softer, then larger differential settlements can be toler-
ated.

The predicted pipe settlements due to thawing are compared with
the allowable settlements based on the structural analysis. Judgments
are made about whether normal burial of the pipe will be acceptable, or
whether deeper burial, abovegrade construction of some kind, or re-
routing to avoid a problem area would be the preferred solution. The
final selection of the design mode may well be based on economic factors, tempered by environmental concerns.

(b) Buoyancy

Large pipelines are buoyant in water at least during their inactive period prior to operation. Oil pipelines will normally not remain
buoyant during operation, whereas gas pipelines will remain buoyant.

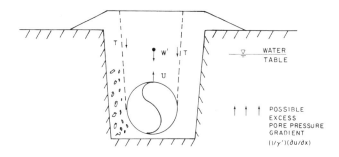

BUOYANCY EQUATION

$$U = W' + 2T$$

where U = net uplift force

        = forces exerted by hydrostatic uplift, plus excess pore pressure gradient, minus the weight of pipe and contents

W' = effective weight of backfill over pipe

T = effective shearing resistance in backfill

Figure 20. Pipe buoyancy design.

If the water table rises (even temporarily) above the top of the pipe, an upward force will be exerted which is calculable from Archimedes' law. The upward force must be resisted by the effective weight of the backfill above the pipe, the shearing resistance developed along shear planes in the backfill, and any additional pipe weights (Fig. 20). The hydrostatic water pressure gradient must be incremented by the excess pore pressure gradient, and this effectively amounts to increasing the density of the fluid used when calculating the upward buoyant force of the pipe. Methods of estimating the excess pore pressure gradient are based on the theory of thaw consolidation, and are detailed earlier.

Consideration of excess pore pressures in pipeline buoyancy design is only necessary in finer, poorly draining soils (i.e. where c_v, the coefficient of consolidation, is less than about 0.04 cm/s).

(c) Lateral Pipe Stability

Low-angle side bends and construction misalignments in the pipe result in out-of-balance forces at the apex of the side bends. In competent unfrozen soils, this load is normally resisted easily by the passive resistance of the soil at the side bend. In permafrost soils, however, lateral pipe stability may be an issue because the soils may have inadequate strength on thawing. Mitigation of this problem may include deeper burial to avoid high-ice-content soils or gravel buttressing to retain the pipe. Concepts of pipe-soil interaction are summarized by Luscher et al. (62).

A field case record of a pipe that was uplifted due to buoyant forces is presented as follows. Near Sans Sault, in the Mackenzie Valley area of the Northwest Territories, an 80-ft (24-m) length of 48-in. (1.2-m) diameter pipe with closed ends was buried in a ditch about 6 ft (1.8 m) deep, through frozen ice-rich peat, and it penetrated an icy organic silt at the ditch bottom. No pipeline weights were used and the ditch was backfilled with the excavated soil, which tended to be frozen peat chunks about 3 to 4 in. (75 to 100 mm) in size. No compaction was used and the spoil was bermed over the ditch. The test was an "inactive" burial and no flow of gas or oil took place. Risers were surveyed on a monthly basis, and no movements were observed initially. Four thaw seasons after installation, the pipe become buoyant and floated to the surface. The eventual flotation was attributed to slow thermal degradation at the site caused by clearing and construction activities.

Well Casings

The installation of production well casings through thick permafrost strata poses a new set of problems for the design engineer. Oil or gas produced from great depths may heat the well casing continuously to temperatures of 50 to 100°C. Thermal simulations show that the permafrost may thaw to a radius of 30 to 65 ft (10 to 20 m) around a single well. The primary concerns arise from elastic deformations in the thawed soil, or possible downward slip of the thawed soil relative to the vertical casing. In either case, depending on soil properties, excessive tractions exerted over a sufficient length of the casing could cause compressive or tensile yield, or buckling of the casing within the permafrost strata.

Several loading mechanisms for the thawing permafrost have been proposed by different researchers (97, 34, 78). Two classes of soil conditions may be distinguished, and two very different consequent casing-soil interactions.

Firstly, where the permafrost soils are initially fairly dense, mostly competent materials having grain-to-grain contact prior to thawing, soil and casing deformations are controlled by the elastic response of the soils in the thaw annulus to various loadings. One of the easier loading mechanisms to understand is that of pore pressure reduction. Assuming the frozen ground is consolidated in equilibrium with a pore ice pressure at depth z of of $\gamma_{ice}z$, then the volumetric contraction associated with the phase transformation of ice to water results in a reduction of the pressure in the pore spaces to a level close to zero. This in turn results in an effective stress increase of

$$\Delta p = \gamma_{ice} \, z$$

This is equivalent to a body force of γ_{ice}, or it may be thought of as an increase in the effective density of the soil. It can be shown from elasticity theory that the downward soil deformations at the center of an infinitely long column of uniform soil are constant with depth, and therefore that the soil strain at the center of the thaw annulus is zero. Assuming the casing remains in intimate contact with

the soil, the work of Palmer (97) can be extended to show that the
casing displacements S for this loading mechanism are of the order of

$$S = \gamma_{ice} \, R^2/4G \qquad (13)$$

where R is the thaw radius and G is the soil shear modulus.

The strain in the casing will be zero only if the soil properties
remain uniform with depth, and the thaw annulus continues downward
indefinitely. Two important factors enter to induce casing strains.
Firstly, nonuniformity of soil properties of layered stratigraphy
causes alternating compression and tension in the casing. These are
predictable from finite element stress analyses, and typical results
for production wells at Prudhoe Bay are provided by Mitchell and
Goodman (78). Potentially more serious are the strains that result as
the casing penetrates the unfrozen, consolidated soil at the base of
the permafrost. A discontinuity results here, as shown on Figure 21.
The thawed soil within the permafrost depth, H, is arching with the
surrounding permafrost, and the vertical effective stress in the thaw
annulus is of the order of

$$\sigma_v' = \gamma' \, (R-a)/2K_o \, \tan \phi' \qquad (14)$$

where γ' = submerged unit weight

K_o = coefficient of lateral earth pressure (0.5 to 1.0) and

a = casing radius

$\tan \phi'$ = friction coefficient for soil (0.4 to 0.8).

This effective stress is very small in comparison with the normally
consolidated vertical effective stress in the underlying unfrozen soil,
$\gamma'H$. An unloading therefore occurs over the circular area where the
thaw annulus intercepts the permafrost base, causing a tensile strain
in the unfrozen soil below the permafrost base, and compression in the
soil and casing above it. The magnitude of these strains can be predicted using relatively complex finite element analyses. Alternatively, fairly straightforward analytical techniques can be used to approximate the maximum tensile strain in the casing just below the permafrost base. From Boussinesq theory for circular footings (see 55), the
maximum vertical strain at a depth of about 0.7 diameter is

$$\xi_{max} = 0.6\Delta p/E_{soil} = 0.6 \, \gamma'H/E_{soil} \qquad (15)$$

where E_{soil} = is the average soil modulus obtained by unloading the
skeleton from its initial overburden pressure to some smaller effective
stress experienced in the thaw annulus.

This simple relationship can be used to predict the maximum casing
strain that might occur at the permafrost base, assuming no slip occurs
between the casing and the soil to relieve the peak strains. (This
analysis can be extended to account for casing slip if required.) More

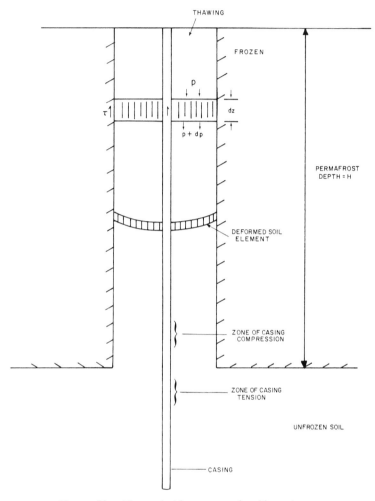

Figure 21.   Thaw subsidence around well casing.

rigorous stress analysis would correctly predict a somewhat smaller
maximum strain than the above, due to the restraining effect of the
edges of the circular loaded area at the permafrost base. The maximum
casing strain is probably about 50-60% of that predicted by the simplified analysis above.

Other loading mechanisms such as the reduction in stiffness of the
soil on thawing have been discussed by Goodman (34), but these may not
be appropriate depending on the geological history of deposition,
freezing and thawing.

The second important casing-soil interaction condition may occur
if the thawed, undrained soil has insufficient shear strength to
support its own weight. In this case, sections of the soil in the thaw
annulus may slip relative to the casing, and this is not accounted for
in the above analysis. If Terzaghi's (117) theory of arching is
applied to the problem, it will be found that the shear support at the
perimeter of the thawed zone, τ, is given by

$$\tau = \gamma'(R-a)/2. \qquad (16)$$

It can be argued that if the shear strength of the soil is less
than $\gamma'(R-a)/2$ for significant depths of the casing, then slip might
occur. To place this in perspective, the soil within a thaw radius of
30 ft (9 m) and having a submerged density of 50 pcf (800 kg/m^3) would
require a minimum shear strength of 750 psf (36 kPa) to have adequate
vertical support. This implies that a soil with more than a few
percent excess ice will likely have inadequate support to prevent
slippage relative to the casing. The implications of this are that a
casing **stress** problem now results, where the casing stresses are calculated by integrating the tractions imposed by the soil along the
casing. For example, a soil with 500 psf (24 kPa) shear strength
slipping relative to an 18-in. (0.46-m) cemented casing (0.5-in. (12.7-
mm) wall) over a vertical depth range of 500 ft (150 m) will induce a
compressive casing stress (0.21% strain) of 62,500 psi (430 MPa). The
possibility of casing buckling should also be examined, and this will
also depend on the soil shear strength, and the lateral support offered
by the soil.

Slope Stability

The freeze-thaw cycle plays a significant role in a variety of
types of landslides associated with the thawing of permafrost slopes in
the Arctic or in the stability of slopes in more temperate regions. A
description of the types of landslides or mass movement forms encountered in thawing slopes is given by McRoberts and Morgenstern (73). A
common mass movement form in the Arctic is referred to as solifluction
(130) and many periglacial researchers have commented on the importance
of the freeze-thaw cycle. Mackay (65) reviewed the frost creep mechanism in which movement is attributed to differential downslope movements
caused by heave occurring perpendicularly to slope and thaw settlement
occurring vertically.

While the process of decrease in shear srength in thawing soils
has long been understood in general terms, thaw-consolidation theory
has now been introduced into slope stability analyses by McRoberts
(72), who extended infinite slope analyses by expressing the factor of
safety F of a slope as

$$F = \frac{c'}{\gamma z \cos\theta \sin\theta} + \frac{\gamma'}{\gamma} \frac{1}{1+2R^2} \frac{\tan\emptyset'}{\tan\theta} \qquad (17)$$

where, γ = bulk density

γ' = submerged or effective unit weight

   c',∅' = the effective cohesion and angle of shearing resistance

      θ = slope angle

      R = thaw consolidation ratio

      z = depth to the thaw front.

If R is zero, we recover the conventional infinite slope equation for
the case of a saturated slope. If R is non-zero, excess pore pressures
are generated during the thaw cycle and contribute to instability,
allowing slopes even with exceedingly flat inclinations to fail.

   A range of techniques is available to analyze thawing slopes and
effect remedial measures. Recent reviews by McRoberts (72) and
Johnston (48) provide good summaries of the methods and techniques
available for permafrost regions.

   Subsoil thawing can also contribute to slope instability in seasonal frost regions. A common occurrence is the failure of cut slopes
for roadways or railroads which are usually cut as steeply as possible.
During the winter months, frost penetrates into the cut slope. During
the spring thaw period, the slope can become saturated by snow melt,
runoff or rain, and by moisture released by thawing of the underlying
frozen ground. Failure of a long, shallow, slab-like form occurs.
Often, by the time the slope is investigated, it is found that the
frost is completely out of the ground.

   An interesting case record of slope instability in permafrost is
available at Inactive Site No. 2 located on the north facing bank of
the Mountain River some 1-1/4 miles upstream from the confluence of the
Mountain and Mackenzie Rivers at Sans Sault Rapids, N.W.T. The slope
itself is compound, with the upper slope inclined at about 9° and the
lower portion at about 16°. Overall the slope is about 200 ft (60 m)
high but the test location was in the higher part of the slope.

   A 48-in. (1.2-m) pipeline was buried in this slope approximately
perpendicular to the contours using conventional "south of 60°" techniques. A ditch some 160 ft (50 m) long was dug during the winter of
1971 on a site previously cleared by dozer. Some 80 ft (24 m) of 48-
inch diameter pipe was installed in the bottom 16° section of the slope
while no pipe was installed in the remainder of the ditch on the 9°
slope. The ditch was then backfilled in the conventional manner with
no extra backfill being imported to make up for the absence of pipe in
the upper part of the ditch. No special provisions were made for
drainage and erosion control. Specially prepared peat boards were
placed on top of the ditch backfill. These boards were impregnated
with seed, which did not germinate.

   The site was visited by a survey party on July 3, 1971 and no
record was made of any instability. The site was revisited on July 17,
1971 by the same personnel and substantial movements had occurred.
Figure 22 is a photograph of the general area on July 31, 1971.

Figure 22.   Failed slope at Inactive Site No. 2, July 31, 1971.

The soils consist of silty clays, clayey silts and silts, classi-fied as CL-CH, CL and ML. High-ice-content soils found at the 4-ft (1.2-m) depth probably represent the top of the permafrost table that would not be thawed under the geothermal conditions obtained prior to site disturbance.

The available evidence strongly indicates that slope instability of the skin flow type (72) occurred, and it is of interest to review the ability to predict such movements. The initial failure had an average depth of 4.0 ft (1.2 m) and based on prefailure data, the frozen bulk density of the soil was in the order of 100 lb/ft^3 (1600 kg/m^3). Peak effective stress parameters determined at low effective normal stresses in drained direct shear are $C' = 0.5$ psi (3.4 kPa) and $\phi' = 30.5°$. Introducing these parameters into equation (17) for a 16° slope angle, we find that, for a factor of safety of 1.0, the R value is predicted to be 0.84, indicating that excess pore pressures contri-buted to the failure. For the 9° slope segment that did not fail, one predicts a factor of safety of 1.26 with the back-calculated R value of 0.84, all other conditions being equal.

The R value predicted by thaw consolidation theory can be ex-pressed in terms of α and c_v. Geothermal studies indicated that $\alpha = 2.4 \times 10^{-2}$ in./s$^{1/2}$ (6.0 $\times 10^{-2}$ cm/s$^{1/2}$) (75), and based on data in Figure 7, a value of $c_v = 1.5 \times 10^{-3}$ cm^2/s is a realistic lower bound. Introducing these values for α and c_v one obtains a predicted R of 0.77, which agrees well with the back-calculated value of 0.84.

Because of the shallow aspect of the failure, the relative influ-ence of effective cohesive, c', in equation (17) is significant. If we assume that the test values for residual strength properties ($c' = 0$, $\phi'_r = 30.3°$) govern, we obtain R = 0.34 back-calculated from that

equation for a 16° slope. Again, there is an indication of excess pore
pressures.

     An R value of about 0.8 at a depth of 4.0 ft (1.2 m), referenced
to original conditions before thaw, signifies an excess pore pressure
of about 1.3 ft (0.4 m) of head above the ground surface. While excess
pore pressures were not measured during the initial failure, subsequent
investigations reported by McRoberts et al. (75) did encounter excess
pore pressure of the order of magnitude of 1.3 ft (0.4 m). Therefore,
it is reasonable to conclude that excess pore pressures generated by
thaw-consolidation processes did contribute to slope instability at
this site.

Embankment Dams

     The near-surface layers in embankment dams in seasonal frost areas
are affected by the freeze-thaw cycle. Protection by means of free-
draining ballast layers on the slopes and thaw-stable soils near the
crest usually will solve the main problems.

     Embankment dams on permafrost present more difficult design
problems. The implications of possible thawing are among the most
critical design considerations affecting the principal requirements for
successful performance of an embankment dam. These requirements
include control of seepage through and beneath the dam, adequate shear
strength in the embankment and foundation to ensure stability against
slides, and sufficient volumetric stability of the foundation and
embankment materials to acceptably restrict both total settlement that
might compromise the crest elevation for storage of flood waters and
differential settlement that might cause cracking of the embankment and
compromise its watertightness and stability against sliding.

     Therefore the designer must first decide whether the nature of the
existing foundation materials and the anticipated characteristics of
embankment materials permit post-construction thawing. If thawing
during operation of the dam and reservoir cannot be permitted (Case I),
then, realizing that water stored in the reservoir is a significant
source of heat tending to cause thawing, the designer chooses among
three alternatives:

     Ia. Remove the offending thaw-unstable foundation materials and
          introduce none in zones of the embankment that will freeze
          and thaw,

     Ib. Pre-thaw the offending thaw-unstable foundation materials and
          introduce none in zones of the embankment that will freeze
          and thaw,

     Ic. Take special measures in design and construction to preserve
          the frozen state in the foundation materials and to create
          and preserve a frozen state in the embankment.

Within the limited number of dams constructed to date in a permafrost
environment, all these alternatives have been applied (43). Thus far
method Ic has been applied only in dams less than about 65 feet (20 m)

in height; while passive methods such as shading of the downstream
slope have been employed, active means of establishing and preserving a
frozen condition involving thermal piles or circulation of cold air
through the embankment and foundation have been found more effective.

If foundation materials are sufficiently thaw-stable or the anticipated distress manifestations deriving from their instability are
judged feasible to control, correct, or counteract (Case II), then
another alternative exists:

IIa. the embankment is configured, and special features included,
to perform acceptably under thawing conditions.

This method also has been applied in dams constructed to date (67, 64,
43), and special design features have included choice of self-healing
embankment materials adaptable to large differential settlements, adoption of flatter embankment slopes, overbuilding the crest elevation to
provide for expected crest settlement, acceleration of thaw consolidation in the foundation by means of sand drains, and provisions in the
design for periodic grouting of foundation materials as operations of
the dam and reservoir cause progressive thawing.

Examples of application of each of the four alternative methods
are cited below. First, however, it is convenient to mention the
obviously essential need for the designer to characterize the properties of the embankment and foundation materials in the unfrozen and
thawing states, and to predict with acceptable confidence the thermal
regime of the embankment and foundation.

Some of the earliest work in the analysis of the thermal regime of
embankment dams was done by Bogoslovskiy (12) who developed analytical
approaches for prediction of steady-state thermal conditions for both
impermeable and pervious dams of homogeneous cross section. At about
the same time approximate methods were being developed independently
for calculation of transient temperature fields in homogeneous and
zoned embankments without seepage; these developments were reported by
Tsytovich et al. (122). These methods and a method reported by
Kudryavstev et al. (51) were applied to several large zoned embankments
under construction in the USSR at that time, but the predicted temperature regimes did not agree well with temperatures measured later during
operation of the dams (43). It was recognized that one of the causes
of the discrepancies is the strong influence of air convection within
rock-fill zones, and an approach to solution of the problem was developed by Mukhetdinov (84). At about the same time a method was reported
by Brown and Johnston (16) for predicting the progression of thawing
beneath pervious and impermeable homogeneous dams, based on simple heat
conduction theory. Kronik and Demin (50) developed a finite element
method of predicting the temperature regime in zoned embankments. A
synthesis of recommended practice for design of dams (129) includes
methods and examples of calculations of the temperature regime within
embankments and their foundaitons. Brief reviews of analytical
approaches were reported by Moisseyev and Moisseyev (79), Kronik (49),
and Tsytovich et al. (123).

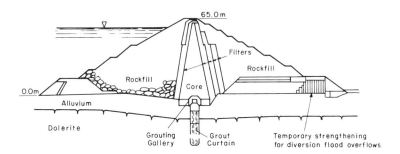

Figure 23. Ust' Khantaika main channel dam (11).

While dikes up to 50 ft (15 m) of great geotechnical interest have been constructed on permafrost in North America in accordance with Case IIa above (64, 67), most dam construction under permafrost conditions has been carried out in the USSR. The effect of the thermal regime on the design of six dams on permafrost in the USSR was reported by Johnson and Sayles (42). In each case the expectation of thawing conditions, or the perceived need to prevent thawing, had an important effect on the design of critical features of the dam. The effects on two of the dams are summarized here.

The Ust' Khantaika hydroelectric project, located north of the Arctic Circle about 90 mi (150 km) south of Noril'sk, includes two sizable embankments, a 210-ft (65-m) main channel dam and a 107-ft (33-m) right saddle dam. The discontinuous or island permafrost is 30 to 60 ft (10 to 20 m) thick. The geotechnical design of the main dam was not unusual (Fig. 23), but conditions during and after construction are worthy of comment. It was intended that the embankment be constructed and operated as a thawed dam. For various reasons, however, the impervious core, partly of glacial moraine soil and partly of gravelly silt, was constructed principally in winter at temperatures down to -40°C (136). Much of the core was constructed of wet silty soil in a semi-liquid state, making its compaction impossible (49). The core became frozen at least in part, despite the specific recommendations of Tsytovich et al. (121) in favor of measures for winter protection of the incomplete core from the development of adverse frozen structure. The project began to be operated in 1975 and the following year gradual thawing of the core and 11.5 ft (3.5 m) of settlement were observed. These developments caused concern (43), and it was also concluded that the voids of the downstream rockfill zone were already completely ice-filled, but that the ice-filled zone did not extend completely through the alluvial gravel subsoil, where seepage was extending the talik on both sides. The buildup of ice in the pore spaces of the downstream rockfill zone, possibly to a condition of complete ice saturation, is a subject of great engineering interest because it illustrates the great effect of free air convection on the thermal regime of rock-fill dams. This effect was discovered earlier at the 250-ft (75-m) high Vilyuy dam, where the expanding zone of frozen, ice-filled rockfill has been accompanied by refreezing of the talik beneath it to a depth of 175 ft (54 m).

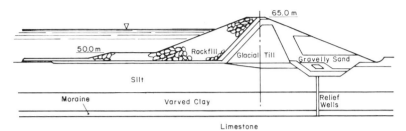

Figure 24. Ust' Khantaika right saddle dam (86).

At the right saddle dam site the subsoils consist of glacial
moraine, and lacustrine-glacial, lacustrine, and bog deposits (86).
The glacial moraines are unsorted mixtures of particles ranging from
cobbles through silt, while the lacustrine and lacustrine-glacial
deposits are clay, varved clay, sandy silt, and silt with sand lenses
and interlayers. Beneath these soils, at a depth of 100 ft (30 m) are
limestone and dolerite rock. The bog deposits, and the semi-liquid
portions of the lacustrine deposits, were removed from the area occu-
pied by the dam. While most of the subsoil was found to be unfrozen,
perennially frozen zones, lenses, and lumps were found in many areas at
temperatures generally around -0.3°C. Wherever these frozen soils were
found with a massive cryogenic structure, they were left in place. But
high-ice-content soils with stratified cryogenic structure were
removed.

Although large lenses of perennially frozen soil were left in the
subsoil, the dam was planned essentially as a thawed dam (86). To
facilitate consolidation of the saturated soils as they thawed, and to
control under-seepage through possible zones of ice-rich soils with a
collapsible structure, a horizontal trench drain was constructed under
the downstream rockfill zone (Fig. 24). Relief wells discharging into
the horizontal drain were installed to tap the more pervious moraine
deposits beneath the varved clay.

The Kolymskaya Hydroelectric Project is now being constructed on
the headwaters of the Kolyma River in the Province of Magadan. The
site is located several hundred kilometers north of the city of
Magadan, near the village of Sinegor'ye. Permafrost is continuous
except for a through talik under the river channel, reaching a depth of
about 1,000 ft (300 m) under the left slope of the valley and ranging
from about 65 ft (20 m) at the right river bank to about 500 ft (150 m)
at the right valley slope. The temperature of the frozen ground, at
the depth of zero seasonal amplitude, is about 18 to 25°F (-4 to -8°C).
The dam will be of rockfill, 413 ft (126 m) in height (Fig. 25).

The dam site is located within a granite batholith. In the lower
part of the more gentle right side, flood plain terraces of diluvial
soil are found to a depth of 65 to 100 ft (20 to 30 m) (33). The soil
characterized by low internal friction and high compressibility upon
thawing and will be completely removed (99). The granite bedrock is
hard in core samples but is intensely fractured. At the surface the

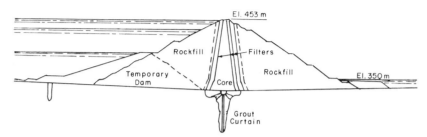

Figure 25. Kolyma Dam (33).

cracks are about 0.8 in. (2 cm) wide and are spaced about 3 or 4 per
meter, becoming narrower and more widely spaced at greater depth (102).
To measure its in-situ permeability the rock first had to be thawed;
for this purpose water from the Kolyma River at 10°C was pumped into
boreholes (99). The rate of thawing obtained in these hydraulic thawing experiments was found to be satisfactory, and subsequent pumping
tests in the thawed rock confirmed the previous belief that the
fissured rock, after thawing, is highly pervious.

With such pervious rock, and the expectation that heat flow from
the reservoir would melt the ice filling the fissures, it was decided
the rock must be prethawed and grouted before filling the reservoir.
This plan is in contrast to the technique used at Vilyuy Dam, where the
less pervious rock was permitted to thaw during operation of the reservoir, with sequential grouting as thawing proceeded. As at Vilyuy, the
grouting will be performed from a concrete grouting gallery directly
beneath the core of the dam. Pre-thawing of the rock will be performed
by pumping in river water, either through the grouting gallery or from
the ground surface.

It was reported that thermal calculations were made to predict the
condition of the core (43). The results, which include effects of
seepage, show that the 0°C isotherm will be in the downstream transition zone. The designers viewed this result as favorable; if, to the
contrary, the core should become frozen it might be brittle and therefore more susceptible to cracking. The calculations showed that the
lower part of the downstream rockfill zone, below tailwater level, will
become ice-filled, but the possible filling of the pores of the upper
part of the rock fill with ice is uncertain. To reduce this possibility, it is intended to try to restrict air convection by covering the
outer slope of the downstream rock-fill zone with broken rock and finer
material. Petrov and Losev (99) reported that thermal studies were
also made with an "electric integrator" (probably an electric analog
computer), which considered heat transfer by seepage and, in an approximate way, included the influence of air convection in the downstream
rockfill zone.

Construction and operation of this large dam, of unprecedented
height among dams on permafrost, will be followed with great interest
as appropriate documentation is presented in the Soviet technical
literature.

Figure 26.   System of cooling pipes exposed along crest
of Irelyakh Dam.

Other dams on permafrost in the USSR of geotechnical interest
described by Johnson and Sayles (42) include the 250-ft (75-m) Vilyuy
Dam (27) already mentioned in connection with free air convection in
rockfill zones. The Irelyakh Dam, a 65-ft (20-m) embankment dam, has
been well described in the literature (e.g. 63), mainly because of the
extensive system of cooling pipes (Fig. 26), installed through the
embankment, cut-off trench, and about 10 ft (3 m) into the foundation,
through which cold ambient air is circulated in the winter to increase
the low temperature reserve and thereby preclude summer thawing.
Anadyr Dam (52) also is of great interest, because of the plan to
install thermal piles for the same purpose served by the air cooling
pipes at the Irelyakh Dam. And finally, development of the design for
the proposed 210-ft (65-m) Ust' Botuobiya Dam will be followed with
interest as it is reported in the technical literature, since early
results from site investigations (42) indicated foundation problems
requiring innovative approaches.

CONCLUDING REMARKS

    In this paper we have sought to identify the principal advances in
geotechnical engineering related to subsoil thawing that have taken
place mainly within the past 1-1/2 decades. A sound theoretical basis
for engineering calculations of the effects of subsoil thawing was
established with the development of a comprehensive one-dimensional
theory of thaw consolidation. With a number of other theoretical
developments that this advance made possible, and with the accumulation
of laboratory and field data on the mechanical properties of thawed
soil, the practitioner can now predict with reasonable confidence the
response of various types of facilities to the thawing of the subsoil.

With this predictive ability the designer now can not only evaluate
alternatives that include preservation of the frozen condition, preservation of thawed condition, and pre-thawing or removal of thaw-unstable
soils, but can also more realistically evaluate design alternatives
that involve thawing subsoils, or periodic cycling between frozen and
thawed states.

We have set out in this paper design approaches and a number of
examples of geotechnical designs for thawing subsoil. We have not
attempted to list the many specific research needs, but the review of
the current technology makes it possible for those researchers among
our readers to identify topics within their areas of interest on which
further studies would be most productive.

ACKNOWLEDGMENTS

    Permission by Inco Metals Co. to use data from Thompson Open Pit
Project is gratefully acknowledged.

APPENDIX 1.--REFERENCES

1. Alkire, B.D. "A Mechanism for Predicting the Effect of Cyclic
    Freeze-Thaw on Soil Behavior," Proceedings 2nd International Sym-
    posium on Ground Freezing, Trondheim, Norway, 1980, pp. 285-296.

2. Anderson, D.M., Pusch, R. and Penner, E. "Physical and Thermal
    Properties of Frozen Ground," Chapter 2, Geotechnical Engineering
    for Cold Regions, O. Andersland and D.M. Anderson (Eds.) McGraw
    Hill, 1978, pp. 37-102.

3. Arctic Construction and Frost Effects Laboratory. "Summary Report
    of Frost Investigations 1944-1947," U.S.A. Cold Regions Research
    and Engineering Laboratory (CRREL) Internal Report 7 (unpub-
    lished), 1950, 34 pp.

4. Auld, R., Roblins, R. and Sangster, R.H. "Foundation Requirements
    in the Canadian Arctic," Proceedings, 24th Canadian Chemical Engi-
    neering Conference, Ottawa, Ontario, October 20-23, 1974, manu-
    script copy, 27 pp.

5. Bagnold, R.A. "The Flow of Cohesionless Grains in Fluid," Philo-
    sophical Translation, Royal Society of London, Series A, Vol. 249,
    1956, pp. 235-297.

6. Barker, W.R. and Brabston, W.N. "Development of a Structural De-
    sign Procedure for Flexible Airport Pavements," U.S. Army Water-
    ways Experiment Station Report No. S-75-17, Vicksburg, Mississip-
    pi, 1975, 261 pp.

7. Barker, W.R. and Parker, Jr., F. "Comparative Performance of
    Structural Layers in Pavement Systems," Vol. IV. Prepared by U.S.
    Army Waterways Experiment Station for Federal Aviation Administra-
    tion, Report No. FAA-RD-73-198, 1977. 143 pp.

8. Bergan, A.T. "Some Considerations in the Design of Asphalt Con-
    crete Pavements for Cold Regions," Ph.D. Dissertation, University
    of California, Berkeley, 333 pp.

9. Bergan, A.T. and Monismith, C.L. "Characterization of Subgrade
    Soils in Cold Regions for Pavement Design Purposes," Highway
    Research Record 431, 1973, pp. 25-37.

10. Bergan, A.T. and Monismith, C.L. "Some Fatigue Considerations in
    the Design of Asphalt Concrete Pavements," Presented at meeting of
    Canadian Technical Asphalt Association, 1972, 59 pp.

11. Biyanov, G.F. "Dams on Permafrost," CRREL Draft Translation 555,
    1976, 234 pp.

12. Bogoslovskiy, P.A. "Investigations on the Temperature Regime of
     Earth Dams Under Permafrost Conditions," Nauchnye Doklady Vysshey
     Shkoly, Stroitel'stvo, No. 1, 1958, pp. 228-238, CRREL Draft
     Translation 22.

13. Broms, B.B. and Yao, L.Y.C. "Shear Strength of Soil After Freez-
     ing and Thawing," Journal, Soil Mechanics and Foundations Divi-
     sion, ASCE, Vol. 90, No. SM4, July, 1964, pp. 1-25.

14. Brown, S.F., Lashine, A.K.F. and Hyde, A.F.L. "Repeated Load
     Triaxial Testing of a Silty Clay," Geotechnique, 25, No. 1, 1975,
     pp. 95-114.

15. Brown, S.F. and Pappin, J.W. "Analysis of Pavements with Granular
     Bases," Transportation Research Record No. 810, 1975, pp. 17-23.

16. Brown, W.G., and Johnston, G.H. "Dikes on Permafrost: Predicting
     Thaw and Settlement," Canadian Geotechnical Journal, Vol. 7, pp.
     365-371.

17. Cedergren/KOA. "Development of Guidelines for the Design of Sub-
     surface Drainage Systems for Highway Pavement Structural
     Sections," Federal Highway Administration, 1972, 24 pp.

18. Chamberlain, E.J. "Overconsolidation Effects of Ground Freezing."
     Engineering Geology, Vol. 18, 1981, pp. 97-100.

19. Chamberlain, E.J. "Identification and Classification of Frost
     Susceptible Soils," Presented at ASCE Spring Convention, Las
     Vegas, Nevada, 1982.

20. Chamberlain, E.J. and S. Blouin. "Densification by Freezing and
     Thawing of Fine Materials Dredged from Waterways," Proceedings,
     3rd International Conference on Permafrost, Edmonton, 1978, pp.
     622-628.

21. Chamberlain, E.J. and Gow, A.J. "Effects of Freezing and Thawing
     on the Permeability and Structure of Soils," Engineering Geology,
     Vol. 13, 1979, pp. 73-92.

22. Chamberlain, E.J. and Cole, D.M. "The Use of Moisture Tension to
     Identify Thaw-Weakened Soil," U.S.A. Cold Region Research and
     Engineering Laboratory, Hanover, N.H., 1983, in preparation.

23. Chamberlain, E.J., Cole, D.M. and Johnson, T.C. "Resilient
     Modulus and Poisson's Ratio for Frozen and Thawed Silt and Clay
     Subgrade Materials," Applications of Soil Dynamics in Cold
     Regions, ASCE Specialty Session, San Francisco, 1977, pp. 229-281.

24. Cole, D.M., Irwin, L.H. and Johnson, T.C. "Effect of Freezing and
     Thawing on Resilient Modulus of a Granular Soil Exhibiting Non-
     linear Behavior," Transportation Research Record No. 809, 1981,
     pp. 19-26.

25. Crory, F.E. "Settlement Associated with Thawing Permafrost,"
     Proceedings, 2nd International Conference on Permafrost, Yakutsk,
     1973, pp. 599-607.

26. Culley, R.W. "Effect of Freeze-Thaw Cycling on Stress-Strain
     Characteristics and Volume Change of a Till Subjected to Repeti-
     tive Loading." Technical Report 13, Saskatchewan Department of
     Highways, 1970, 73 pp.

27. Demidov, A.N.   "Dam in the Vilyuy Hydraulic Power System," Trudy
     Gidroproekta, No. 34, 1973, pp. 64-77. CRREL Draft Translation
     526.

28. Duncan, J.M., Monismith, C.L. and Wilson, E.L. "Finite Element
     Analyses of Pavements," Highway Research Record No. 228, 1968, pp.
     18-33.

29. Esch, D. "Thawing of Permafrost by Passive Solar Methods," Pro-
     ceedings, 4th Canadian Permafrost Conference, Calgary, 1981.

30. Fredlund, D.G. "Appropriate Concepts and Technology for Unsatu-
     rated Soils," Canadian Geotechnical Journal, Vol. 16, 1979, pp.
     121-139.

31. Fredlund, D.G., Bergan, A.T. and Sauer, E.K. "Deformation Charac-
     terization of Subgrade Soils for Highways and Runways in Northern
     Environments," Canadian Geotechnical Journal, Vol. 12, No. 2,
     1975, pp. 213-223.

32. Fredlund, D.G., Morgenstern, N.R. and Widger, W.A. "The Shear
     Strength of Unsaturated Soils," Canadian Geotechnical Journal,
     Vol. 15, 1978, pp. 313-321.

33. Gluskin, Ya.E., Losev, E.D., Petrov, V.G. and Frumkin, N.V.
     "Kolyma Hydroelectric Station," Hydrotechnical Construction, No.
     8, August, Translation for ASCE, 1974, pp. 722-727.

34. Goodman, M. "Loading Mechanisms in Thawed Permafrost Around
     Arctic Wells," Journal of Pressure Vessel Technology, Vol. 99,
     No. 4, 1977, pp. 641-645.

35. Harlan, R.L. and Nixon, J.F. "Chapter 3, Ground Thermal Regime,"
     Geotechnical Engineering in Cold Regions. D. Anderson and O.
     Andersland (Eds.), pp. 103-163, McGraw-Hill, N.Y., 1978.

36. Inco. "Geotechnical Data Report," Hardy Associates (1978) Ltd.
     report. Thompson Open Pit, Inco Metals Company, a unit of Inco
     Ltd., 1981.

37. Ingersoll, J. "Moisture Retention Characteristics of Soils Using
     Tempe Pressure Sells," CRREL Technical Note (unpublished), 1976.

38. Jahns, H., Miller, T., Power, L., Rickey, W., Taylor, T. and
     Wheeler, J. "Permafrost Protection of Pipelines," Proceedings,
     2nd International Permafrost Conference, Yakutsk, 1973, pp. 673-
     684.

39. Jessberger, H.L., and Carbee, D.L. "Influence of Frost Action on
    the Bearing Capacity of Soils," Highway Research Record 304, 1970,
    pp. 14-26.

40. Jessberger, H.L., "Frost Susceptibility Criteria," Highway
    Research Record 429, 1973, pp. 40-46.

41. Johnson, T.C., "Is Graded Aggregate Base the Solution in Frost
    Areas?" Proceedings, Conference on Utilization of Graded Aggregate
    Base, National Crushed Stone Association, 1974, pp. IV-1 to IV-18.

42. Johnson, T.C. and Sayles, F.H. "Effect of Ground Thermal Regime
    on Designs of Embankment Dams on Permafrost in the USSR," CRREL
    Technical Note (unpublished), 1979, 38 pp.

43. Johnson, T.C. and Sayles, F.H. "Embankment Dams on Permafrost in
    the USSR," CRREL Special Report 80-41, 1980, 59 pp.

44. Johnson, T.C., Berg, R.L., Carey, K.L. and Kaplar, C.W. "Roadway
    Design in Seasonal Frost Areas," Transportation Research Board
    Synthesis of Highway Practice, No. 26, 1975, 104 pp.

45. Johnson, T.C., Cole, D.M. and Chamberlain, E.J. "Influence of
    Freezing and Thawing on the Resilient Properties of a Silt Soil
    Beneath an Asphalt Concrete Pavement," CRREL Report 78-23, 1978,
    49 pp.

46. Johnson, T.C., Cole, D.M. and Chamberlain, E.J. "Effect of
    Freeze-Thaw Cycles on Resilient Properties of Fine-Grained Soils,"
    Proceedings, 1st International Symposium on Ground Freezing,
    Bochum, 1978, p 247-276.

47. Johnson, T.C., Cole, D.M. and Irwin, L.H. "Characterization of
    Freeze/Thaw-Affected Granular Soils for Pavement Evaluation,"
    Paper submitted to 5th International Conference on Structural
    Design of Asphalt Pavements, Amsterdam, 1982.

48. Johnston, H. (Ed.) Permafrost - Engineering Design and Construc-
    tion. J. Wiley and Sons, N.Y., 1981, 540 pp.

49. Kronik, Ya.A. "Hydrotechnical Construction," Vol. 3, Chapter 14,
    Engineering Geology of the USSR, G.A. Golodkovskoy, Ed., Moscow
    State University (in Russian), pp. 254-270.

50. Kronik, Ya.A., and Demin, I.I. "Calculation of the Temperature
    Regime of Dams of Local Materials by the Method of Finite
    Elements," Gidrotekhicheskoe Stroitel'stvo, No. 2, 1979, pp.
    26-30.

51. Kudryavtsev, V.A., Garagulya, L.C., Kondratyeva, K.A. and
    Melamyed, V.G. "Fundamentals of Frost Forecasting in Geological
    Engineering Investigations," CRREL Draft Translation 606, 1974,
    489 pp.

52. Kuznetsov, A.L. "Dam of the Anadyrsk Thermal Electric Power

Station," Trudy Gidroproyekta, No. 34, pp. 88-100. CRREL Draft
Translation 659, 1973.

53. Lachenbruch, A.H. "Periodic Heat Flow in a Stratified Medium with
Application in Permafrost Problems," U.S. Geological Survey
Bulletin 1052, pp. 51, 1959.

54. Lachenbruch, A.H. "Some Estimates of the Thermal Effects of a
Heated Pipeline in Permafrost," U.S. Geological Survey Circular
632, Vol. 5, 1970, 23 pp.

55. Lambe, T. and Whitman, R. Soil Mechanics, J. Wiley and Sons,
N.Y., 1969.

56. Linell, K.A., Hennion, F.B. and Lobacz, E.F. "Corps of Engineers'
Pavement Design in Areas of Seasonal Frost," Highway Research
Record No. 33, 1963, pp. 76-136.

57. Linell, K. "Long-Term Effects of Vegetation Cover on Permafrost
Stability in an Area of Discontinuous Permafrost," Proceedings,
2nd International Permafrost Conference, Yakutsk, 1973, pp.
688-693.

58. Linell, K. and Lobacz, E., "Design and Construction of Foundations
in Areas of Deep Seasonal Frost and Permafrost," CRREL Special
Report 80-34, 310 pp.

59. Lobacz, E. and Quinn W. "Thermal Regime Beneath Buildings Constructed on Permafrost," Proceedings, 1st Permafrost Conference,
Lafayette, Indiana, NAS/NRC Publication 1287, 1963, pp. 247-252.

60. Lovering, W.R. and Cedergren, H.R., "Structural Section Drainage,"
Proceedings, International Conference on Structural Design of
Asphalt Pavements, University of Michigan, 1962, pp. 773-784.

61. Luscher, U. and Afifi, S.S. "Thaw Consolidation of Alaskan Silts
and Granular Soils," Proceedings, 2nd International Conference on
Permafrost, Yakutsk, 1973, pp. 325-333.

62. Luscher, U., Thomas, H.P. and Maple, J.A. "Pipe-Soil Interaction,
Trans-Alaska Pipeline," Proceedings, ASCE Specialty Conference on
Pipelines in Adverse Environments, New Orleans, January, 1979, pp.
486-502.

63. Lyskanov, G.A. "Experience in Building a Frozen Type Dam in
Yakutiya," Yakutsk, Iakutskoe khizhnoe izd-vo, 70 pp. CRREL Draft
Translation TL 479, 1964.

64. MacDonald, D.H. "Design of Kelsey Dikes," Proceedings, 1st International Conference on Permafrost, Purdue University, Lafayette,
Indiana, 1973, pp. 492-496.

65. Mackay, J.R. "Active Layer Slope Movement in a Continuous Permafrost Environment, Garry Island, Northwest Territories," Canadian
Journal of Earth Sciences, Vol. 18, 1981, pp. 1666-1680.

66. MacLeod, D.R. "Some Fatigue Considerations in the Design of Thin Pavements," M.Sc. Thesis, University of Saskatchewan, Saskatoon, 1971.

67. Macpherson, J.G., Watson, G.H. and Koropatnick, A. "Dikes on Permafrost Foundation in Northern Manitoba," Canadian Geotechnical Journal, Vol. 7, 1970, pp. 356-364.

68. Malyshev, M.A. "Deformation of Clays During Freezing and Thawing," CRREL Draft Translation 338, 1969, 6 pp.

69. McDougall, J. "The Beaufort Gas Project Surface Facilities," Proceedings, 2nd International Symposium on Cold Regions Engineering, University of Alaska, 1976, pp. 383-400.

70. McGowan, A. and Radwan, A.M. "The Presence and Influence of Fissures in the Boulder Clays of West Central Scotland," Canadian Geotechnical Journal, Vol. 12, No. 11, 1975, pp. 84-97.

71. McKim, H.L., Berg, R.L., McGaw, R.W., Atkins, R.T. and Ingersoll, J. "Development of a Remote-Reading Tensiometer/Transducer System for Use in Subfreezing Temperatures," Proceedings, 2nd Conference on Soil-Water Problems in Cold Regions, Edmonton, Alberta, 1976, pp. 31-45.

72. McRoberts, E.C. "Stability of Slopes in Permafrost," Ph.D. Thesis, University of Alberta, Edmonton, 1973.

73. McRoberts, E.C. and Morgenstern, N.R. "The Stability of Thawing Slopes," Canadian Geotechnical Journal, Vol. 11, 1974, pp. 447-469.

74. McRoberts, E.C. and Nixon, J.F. "A Theory of Soil Sedimentation," Canadian Geotechnical Journal, Vol. 13, 1976, pp. 294-309.

75. McRoberts, E.C., Fletcher, E.B. and Nixon, J.F. "Thaw Consolidation Effects in Degrading Permafrost," Proceedings, 3rd International Conference on Permafrost, Edmonton, Vol. 1, 1978, pp. 693-699.

76. McRoberts, E.C., Law, T.C. and Moniz, E. "Thaw Settlement Studies in the Discontinuous Permafrost Zone," Proceedings, 3rd International Conference on Permafrost, Edmonton, Vol. 1, 1978, pp. 700-706.

77. Mickleborough, B.W. "An Experimental Study of the Effects of Freezing on Clay Subgrades," M.Sc. Thesis, University of Saskatchewan, Saskatoon, 1970, 219 pp.

78. Mitchell, R. and Goodman, M. "Permafrost Thaw-Subsidence Casing Design," ASME paper 76-SPE-6060, 1976.

79. Moisseyev, S.N. and Moisseyev, I.S. Kamyeno-Zelyanyye Plotiny (Rock-Earth Dams) (in Russian), Energhiya Publishing House, Moscow, 1977, 280 pp.

80. Morgenstern, N.R. "Geotechnical Engineering and Frontier Resource Development," Geotechnique, Vol. 33, No. 3, 1981, pp. 305-365.

81. Morgenstern, N.R. and Nixon J.F. "One-Dimensional Consolidation of Thawing Soils," Canadian Geotechnical Journal, Vol. 8, 1971, pp. 558-565.

82. Morgenstern, N.R. and Nixon, J.F. "An Analysis of the Performance of a Warm-Oil Pipeline in Permafrost," Canadian Geotechnical Journal, Vol. 12, 1975, pp. 199-208.

83. Morgenstern, N.R. and Smith, L.B. "Thaw Consolidation Tests on Remolded Clays," Canadian Geotechnical Journal, Vol. 10, 1973, pp. 25-40.

84. Mukhetdinov, N.A. "Thermal Regime of the Downstream Shoulder of Rockfill Dams," Leningrad, VNIIG Izvestiya, Vol. 90, pp. 275-294, CRREL Draft Translation 586, 1969.

85. Murfitt, A.W., McMullen, W.B., Baker, M. and McPhail, J.F. "Design and Construction of Roads on Muskeg in Arctic and Sub-arctic Regions," Proceedings, 16th Muskeg Research Conference, National Research Council of Canada, Ottawa, 1975, pp. 152-185.

86. Myznikov, Yu.N., Zhilenas, S.V. and Ten, N.A. "Construction of Right-Bank Dam at Ust'-Khantaisk Hydroelectric Plant," Hydrotechnical Construction, No. 4, April, translation for ASCE, 1975, pp. 309-313.

87. Nixon, J.F. "Geothermal Aspects of Ventilated Pad Design," Proceedings, 3rd International Conference on Permafrost, Edmonton, Vol. 1, 1978, pp. 840-846.

88. Nixon, J.F. "Some Aspects of Road and Airstrip Pad Design in Permafrost Areas," Canadian Geotechnical Journal, Vol. 16, 1979, pp. 222-225.

89. Nixon, J.F. and Halliwell, D. "Practical Applications of a Versatile Geothermal Simulator," Presented at ASME winter meeting, Phoenix, Arizona, 1982.

90. Nixon, J.F. and Hanna, A.J. "The Undrained Strength of Some Thawed Permafrost Soils," Canadian Geotechnical Journal, Vol. 16, 1979, pp. 420-427.

91. Nixon, J.F. and Ladanyi, B. "Thaw Consolidation," Geotechnical Engineering for Cold Regions, Chapter 4, O. Andersland and D.M. Anderson (Eds.), McGraw-Hill, N.Y., 1978, pp. 164-215.

92. Nixon, J.F. and McRoberts, E.C. "A Design Approach for Pile Foundations in Permafrost," Canadian Geotechnical Journal, vol. 13, 1976, pp. 40-50.

93. Nixon, J.F. and Morgenstern, N.R. "The Residual Stress in Thawing Soils," Canadian Geotechnical Journal, vol. 10, 1973, pp. 571-580.

94. Nixon, J.F. and Morgenstern, N.R. "Thaw Consolidation Tests on Undisturbed Fine-Grained Permafrost," Canadian Geotechnical Journal, Vol. 11, 1974, pp. 202-214.

95. Pagen, C.A. and Khosla, V.K. "Frost Action Characteristics of Compacted Cohesive Soil," Ohio State University Report No. EES 248-4, 1968.

96. Painter, L.J. "Analysis of AASHO Road Test Asphalt Pavement Data by the Asphalt Institute," Highway Reseach Record No. 71, 1965, pp. 15-38.

97. Palmer, A.C. "Thawing and Differential Settlement of Ground Around Oil Wells in Permafrost," Proceedings, 2nd International Conference on Permafrost, USSR Contribution, Yakutsk, 1972, pp. 619-624.

98. Paterson, T.T. "The Effects of Frost Action and Solifluction Around Baffin Bay and in the Cambridge District," Geological Society, London, Q.J., Vol. 96, Part 1, 1940, pp. 99-130.

99. Petrov, V.G. and Losev, E.D. "Dam of the Kolyma Hydroelectric Power Plant," Trudy Gidroproekta, No. 34, pp. 78-87. CRREL Draft Translation TL 563, 1973.

100. Phukan, A. and Andersland, O. "Foundations for Cold Regions," Geotechnical Engineering for Cold Regions, Chapter 6, Andersland and Anderson (Eds.), McGraw-Hill, pp. 276-362.

101. Plyat, Sh.N., Mukhetdinov, N.A. and Smirnov, E.A. "Thermal Regime of Earth-Rock Dams Constructed in the Far North," Leningrad, Soviet-American Working Seminar, Technology of Construction of Structures Under Cold Climate Conditions, May 25-26. CRREL Draft Translation 658, 1977, 14 pp.

102. Pogrebiskiy, M.I. and Chernyshev, S.N. "Determination of the Permeability of the Frozen Fissured Rock Massif in the Vicinity of the Kolyma Hydroelectric Power Station," CRREL Draft Translation 634, 1975, 13 pp.

103. Pretorius, P.C. "Design Considerations for Pavements Containing Soil-Cement Bases," Ph.D. Dissertation, University of California, Berkeley, California, 1969.

104. Quinn, W.F., Carbee, D. and Johnson, T.C. "Membrane Encapsulated Soil Layers (MESL) for Road Construction in Cold Regions," Proceedings, OECD Symposium on Frost Action on Roads, Oslo, Norway, Vol. II, 1973, pp. 417-438.

105. Raad, L. and Figueroa, J.L. "Load Response of Transportation Support Systems," ASCE Transportation Engineering Journal, TE1, 1980, pp. 111-128.

106. Robnett, Q.L. and Thompson, M.R. "Effect of Lime Treatment on the

Resilient Behavior of Fine-Grained Soils," Transportation Research Board Record 560, 1976, pp. 11-20.

107. Roggensack, W.D. "Geotechnical Properties of Fine-Grained Permafrost Soils," Ph.D. Thesis, University of Alberta, Edmonton, 1977.

108. Rowe, P.W. "The Relevance of Soil Fabric to Site Investigation Practice," Geotechnique, Vol. 22, No. 2, 1972, pp. 195-300.

109. Sage, J.S. and D'Andrea, R.A. Personal communication, Worcester Polytechnic Institute, Department of Civil Engineering, 1981.

110. Sanger, F. "Foundation for Structures in Cold Regions," CRREL Monograph III-C4, 1969, 91 pp.

111. Seed, H.B., Chan, C.K. and Monismith, C.L. "Effects of Repeated Loading on the Strength and Deformation of Compacted Clay," Proceedings, Highway Research Board, Vol. 34, 1955, pp. 541-558.

112. Shusherina, Y.P. "Variation of Physio-Mechanical Properties of Soils Under the Action of Cyclic Freeze-Thaw," CRREL Draft Translation 255, 1971, 11 pp.

113. Smith, N. "Construction and Performance of Membrane Encapsulated Soil Layers in Alaska," CRREL Report 79-16, 27 pp.

114. Speer, T.L., Watson, G.H. and Rowley, R.K. "Effects of Ground-Ice Variability and Resulting Thaw Settlements on Buried Warm-Oil Pipelines," North American Contribution, 2nd International Conference on Permafrost, Yakutsk, 1973, pp. 746-752.

115. Sykes, J.F., Lennox, C. and Charlwood, R.G. "Finite Element Permafrost Thaw Settlement Model," ASCE, Journal of Geotechnical Engineering Division, Vol. 100, 1974, pp. 1185-1201.

116. Taber, S. "Perennially Frozen Ground in Alaska," Geological Society of America Bulletin, Vol. 54, 1943, pp. 1435-1548.

117. Terzaghi, K. Theoretical Soil Mechanics, J. Wiley and Sons, N.Y., 1943, 510 pp.

118. Thomson, S., and Lobacz, E.F. "Shear Strength at a Thaw Interface," North American Contribution, 2nd International Permafrost Conference, National Academy of Sciences, 1973, pp. 419-426.

119. Tsytovich, N.A. The Mechanics of Frozen Ground. McGraw-Hill, N.Y., 1975, 426 pp.

120. Tsytovich, N.A., Zaretskii, Y.K., Grigoreva, V.G. and Termartrosyan, Z. "Consolidation of Thawing Soil," Proceedings, 6th International Conference on Soil Mechanics and Foundation Engineering, No. 1, 1965, pp. 390-394.

121. Tsytovich, N.A., Kronik, Ya.A., Loseva, S.G. and Nozbran, V.F. "Studies of the Temperature-Moisture Regime of the Soil in the

Core of the Khantay Dam (in Russian)," Nauchnyye issledovaniya po gidrotekhnike v 1971 g. (Scientific Studies on Hydraulic Engineering in 1971), Leningrad, VNIIG, 1972.

122. Tsytovich, N.A., Ukhova, N.V. and Ukhov, S.B. "Prediction of the Temperature Stability of Dams Built of Local Materials on Permafrost Bases." Leningrad, Stroizdat, CRREL Draft Translation 435, 1972, 153 pp.

123. Tsytovich, N.A., Kronik Ya.A. and Biyanov, G.F. "Design, Construction and Operation of Earth Dams in Arctic Regions and Under Permafrost Conditions," Proceedings, 3rd International Conference on Permafrost, Vol. 2, 1979, pp. 137-149.

124. Udd, J.E. and Yap, S.M. "Strength Reductions Due to the Thawing of Frozen Ores," 2nd International Symposium of Ground Freezing, Trondheim, Norway, 1980, pp. 309-324.

125. U.S. Department of the Army "Calculation Methods for Determination of Depths of Freeze and Thaw in Soils," Technical Manual TM-852-6, 1966, 89 pp.

126. USSR Academy of Sciences "Thawing Ground as Subsoil for Structures (Ottaivaiushiye grunty kak osnovanyiye sooruzhenyiy)," Scientific Council on Frozen Ground, USSR Academy of Sciences, Moscow (in Russian), 1981, 96 pp.

127. "U.S.S.R. Foundation Handbook," National Research Council of Canada Technical Translation No. 1865, by V. Poppe, 1976.

128. Vallejo, L.E. "Stability of Thawing Slopes," International Conference on Soil Mechanics and Foundation Engineering, Stockholm, Paper 11/44, 1981.

129. VNII VODGEO (All-Union Scientific Research Institute for Water Supply, Channelization, Hydrotechnical Structures, and Engineering Geology), L.N. Kuz'mina, Ed. "Recommended Practice for the Design and Construction of Earth Dams for Industrial and Potable Water Supply in the Far North and Permafrost Areas," CRREL Draft Translation 647, 118 pp.

130. Washburn, A.L. "Periglacial Processes and Environments," Arnold, London, 1973, 320 pp.

131. Waters, E. "Heat Pipes to Stabilize Piles on Elevated Alaska Pipeline Sections," Pipeline Oil and Gas Journal, 1974, August, pp. 46-58.

132. Watson, G.H., Slusarchuk, W.A. and Rowley, R. "Determination of Some Frozen and Thawed Properties of Permafrost Soils," Canadian Geotechnical Journal, Vol. 10, 1973, pp. 592-606.

133. Xiaobai, C., Ping, J. and Yaging, W. "Some Characteristics of Water Saturated Gravel During Freezing and Its Application,"

    Proceedings, 2nd International Symposium on Ground Freezing,
    Norwegian Institute of Technology, Trondheim, Norway, 1988, pp.
    692-712.

134. Yao, L.Y.C. and Broms, B.B. "Excess Pore Pressures Which Develop
     During Thawing of Fine-Grained Subgrade Soils," Highway Research
     Record, No. 101, pp. 39-57.

135. Zaretskii, Y.K. "Calculations of the Settlement of Thawing Soil,"
     Soil Mechanics and Foundation Engineering, No. 3, 1968, pp.
     151-155.

136. Zhilenas, S.V., Myznikov, Yu.N. and Ten, N.A. "Winter Placement
     of Moraine Materials in the Dam of the Ust'-Khantaisk Hydro
     Development," Hydrotechnical Construction, No. 3, March, Transla-
     tion for ASCE, pp. 23-26.

SURVEY OF METHODS FOR CLASSIFYING FROST SUSCEPTIBILITY

By Edwin J. Chamberlain,[1] M. ASCE, Paul N. Gaskin,[2]
David Esch,[3] M. ASCE and Richard L. Berg,[1] M. ASCE

ABSTRACT

Methods for determining the frost susceptibility of soil and granular materials used in road and airfield construction are reviewed. The
methods employed by transportation departments in the United States,
Canada and Europe are included. Three levels of classification are
identified -- Type I, which is based on a specified particle size; Type
II, which is generally based on soil type; and Type III, which requires
a laboratory freezing test. Two critical particle sizes appear frequently, 0.074 and 0.020 mm. The most common basis is the Casagrande
criteria; however, few transportation agencies use the same method, as
modifications have been made to address specific problems. The reliability of most criteria is uncertain because few have been rigorously
validated. Transportation agencies should have all three types available, and possibly a fourth, even more discriminating method, to select
criteria appropriate to the task. The criteria can be narrowly focused
on regional problems of state or provincial agencies or more widely
based on problems of national agencies. The performance of the criteria
should be documented to determine the reliability.

INTRODUCTION

Determining the frost susceptibility of soils and granular base
materials is an important factor to highway engineers in regions of seasonal frost. The search for reliable methods to evaluate the frost susceptibility of soils has gone on for at least the past 50 years. More
than 100 methods have been proposed since Taber's (31) treatise on the
mechanism of ice segregation in soils and Casagrande's (4) conclusion
that "under natural freezing conditions and with sufficient water supply
one should expect considerable ice segregation in non-uniform soils containing more than three percent of grains smaller than 0.02 mm, and in
very uniform soils containing more than 10 percent smaller than 0.02
mm."

Earlier reports by Chamberlain (6,7) attempted to identify all
frost susceptibility criteria and testing methods reported in the literature. As a result, a perspective on the origins and development of
frost suceptibility tests was developed. Many of the methods reviewed
in these earlier reports have never been used, and thus, their utility

[1] Research Civil Engineer, U.S. Army Cold Regions Research and Engineering
Laboratory (CRREL), Hanover, New Hampshire.
[2] Associate Professor, Department of Civil Engineering, Queen's University,
Kingston, Ontario, Canada.
[3] Chief of Highways Research, State of Alaska Department of Transportation
and Public Facilities, Anchorage, Alaska.

remains uncertain. Others have fallen into disuse or have evolved into
more modern methods. This report will not dwell on unused or outdated
methods, but will concentrate on frost susceptibility criteria and testing methods currently in use or being considered for use by state, provincial and national agencies as compliance requirements for the design
of roads and/or airfields in seasonal frost regions.

This report relies heavily on the earlier reviews by Chamberlain
(6, 7) and a recent special issue of Frost i Jord (16) published by the
Norwegian Committee on Permafrost. In addition, information from state
and national agencies in North America has been updated by polling the
individual agencies. The period of time surveyed reaches back to the
early work of Taber (31), Casagrande (4), Beskow (2) and Ducker (11) and
includes methods reported through May 1982. Although attempts were made
to identify all methods presently in use, some may have been missed.
The most serious omissions may be from the Asian and eastern European
nations because of the difficulty in gaining access to their literature.

TYPES OF FROST SUSCEPTIBILITY TESTS

The surveys by Chamberlain (6, 7) identified six fundamentally different categories of laboratory tests for determining the frost susceptibility of soils and granular base materials. They included tests
based on: 1) particle size or soil classification, 2) pore size, 3)
interaction of soil and water, 4) frost heave, 5) thaw weakening, and 6)
freezing without frost heave or thaw weakening.

Of these tests, only those based on particle size or soil classification, frost heave, and thaw weakening have been adopted for use as
compliance requirements for determining the frost susceptibility of
soils and granular materials.

The pore size characteristic methods, the soil-water interaction
methods (such as those based on the moisture retention curve, capillary
rise, unsaturated permeability or centrifuge moisture content), and the
soil freezing tests (based on the heave pressure or pore water suction
during freezing or the hydraulic conductivity of the partially frozen
zone beneath a growing ice lens) have not yet been adopted for use by
any agency nor are any being considered for adoption. They remain the
subject of research and experimental application.

FROST SUSCEPTIBILITY CLASSIFICATION LEVELS

During this review, we have found that the frost susceptibility of
soils and granular materials used in road and airfield construction is
generally estimated at three levels of sophistication (Table 1). The
first level -- Type I -- is primarily based on the percent finer than a
specified particle size, commonly 0.074 mm (standard USC #200 sieve) or
0.02 mm. In some cases, a uniformity designation is an added requirement.

The second level -- Type II -- is generally based on soil type or
classification or particle size curves, and supplementary tests for

Table 1. Levels of Frost Susceptibility Testing.

Test Level	Types of Information Required
Type I	Specified particle size, uniformity[1]
Type II	Soil type or classification, Atterberg limits, capillarity, permeability, minerology, uniformity coefficient, sand equivalent, moisture conditions, density, saturated CBR, etc.
Type III	Frost heave, heave rate, thaw CBR, etc.; laboratory freezing test data

[1] Uniformity coefficient C_u or other qualitative uniformity descriptor.

Atterberg limits, capillarity, or permeability. Type II methods require a more complete description of soil characteristics than Type I methods and are also more time consuming and costly.

The third level — Type III — requires a laboratory freezing test and observations of frost heave and/or thaw weakening. Freezing tests require special laboratory equipment and considerable time and money to conduct. Because of these added time and funding constraints they are not widely used, particularly within the United States and Canada.

FROST SUSCEPTIBILITY CLASSIFICATION METHODS

United States - State Departments of Transportation

Table 2 summarizes the frost susceptibility classification methods employed in the United States. Only 22 states have compliance requirements for frost susceptibility. Figure 1 shows the distribution of the states requiring the identification of frost susceptible materials for road construction purposes. As expected, most of these states lie in the northern regions, the only exception being Arizona, which has many miles of roads in cold mountainous areas.

Of the states employing frost susceptibility criteria, seven specify only the very simple Type I criteria and fifteen require the more complicated Type II criteria. Six employ both types. None of the states canvassed require a laboratory freezing test. However, Alaska, New Hampshire and Michigan have considered including laboratory frost heave tests in their frost susceptibility compliance requirements. Michigan specifies the U.S. Army Corps of Engineers CRREL frost heave test (8, 24), but restricts its use to special research studies because of the time required to conduct the test. New Hampshire has developed its own laboratory freezing test (36), but restricts its use to special studies because of insufficient field validation. Alaska has also developed its own freezing test, but with the exception of research projects, has restricted its use to stabilized soils.

Table 2. Summary of Frost Susceptibility Classification
Methods in the United States.

State or Agency	Type of Classification[1]	Allowable Amount (%) finer than 0.074 mm	0.02 mm	Allowable Plasticity Index (%)[2]	Other Restrictions
Alaska	II,(III)	6-100	--	--	Overburden depth, frost heave test
Arizona	I, II	--	3, 10	x	Soil classification
Connecticut	II	10	3, 10	0	Uniformity
Illinois	I, II	--	3, 10	x	Soil classification
Indiana	I, II	8	3, 10	x	Soil classification
Iowa	II	15	--	x	Soil class., organic content, Proctor density
Maine	I, II	--	3, 10	x	Soil classification
Massachusetts	I	8-12	--	--	--
Michigan	II, (III)	7-10	--	--	Pedological class., drainage test, frost heave test
Minnesota	II	7-15	--	--	Textural class., moisture cond.
Montana	I	--	3, 10	--	Uniformity
New Hampshire	I,[III]	10-12	--	--	Frost heave test
New York	I	10	--	--	--
North Dakota	I	15	--	--	Percent silt
Ohio	II	50	--	>10	
Oregon	II	8	--	<6	Sand equivalent, liquid limit
Pennsylvania	I, II	--	3, 10	x	Soil class
Rhode Island	I	--	1	--	Uniformity
Vermont	I	8-15	--	--	--
Washington	II	10	--	--	Sand equivalent
West Virginia	I, II	--	3, 10	--	Soil classification
Wisconsin	II	2-15	--	--	Pedological class., water table
Asphalt Institute	I	7	--	--	--
Casagrande	I	--	3, 10	--	Uniformity
National Crushed Stone Assoc.	{III}	--	--	--	Frost heave test
U.S. Army Corps of Engineers	I,II,III	--	1.5, 3	x	Soil classification, frost heave test
U.S. Dept of Transportation	I	5-11	--	--	--

Note: 1. ( ) designates test used for research only
[ ] designates test under consideration but not adopted,
{ } designates test abandoned.
2. x designates that the plasticity index is required; the
application is given in the text.

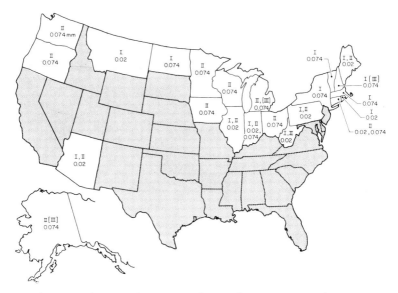

Figure 1. Distribution of states employing frost susceptibility criteria and test levels employed (I, II, and/or III) and critical particle
sizes (0.02 and/or 0.074 mm). [ ] indicates test not in routine use.

Figure 1 also summarizes the critical particle size for each of the
frost susceptibility classification methods. It should be noted that 7
specify only 0.02 mm, while 12 require only 0.074 mm. Two specify both
of these particle sizes.

The 0.074 mm critical particle size has probably been selected by a
larger number of agencies because it is the lower limit of the particle
size distribution as determined by the sieve analysis. This test is
normally conducted for road and airfield construction projects. It is
convenient to use and does not burden projects with additional testing,
whereas the 0.020 mm particle size criterion requires the addition of
the hydrometer and specific gravity tests to the material testing program. The latter tests are not always required by road and airfield
construction agencies.

Michigan uses a visual pedological soil classification system to
determine the frost susceptibility of soils in addition to particle size
limitations, and supplements this method with a laboratory drainage test
for granular base and subbase materials.

Not all of the states have frost susceptibility criteria for all
soil materials in pavement systems. Three specify criteria for subgrade
materials only and assume that the standard specifications for base and
subbase materials exclude frost susceptible materials. One state has
frost susceptibility criteria for base and subbase materials only. A

Table 3. Summary of Freezing Tests.

Country/State/ Province	Confinement method	Freezing Mode[1]	Cold Plate Temperature[2] (°F)	Surcharge[3] (psi)	Number of Freeze- Thaw Cycles	Test Duration (days)	Critical Frost Susceptibility Factors[4] Frost Heave	Thaw Weakening
USA/Alaska [5] [6]	Plexiglass rings	A	14	0.5	1	3	h_r	none
Austria	Plexiglass rings	A	5	0.7	2-4	16-21	h	CBR_T
Canada/British Columbia	Plastic film	A	1.5	0.25	1	10	h	none
Denmark[5]	None	A	1.5	≈0	1	10	h	CBR_T
Federal Republic[7] of Germany	Tapered plexiglass	A	-0.5	0.85	7	7	h,h_r,h_p	CBR_T
France	Foam rubber tube	A	22	0	1	6-8	p	none
German Democratic Republic[8] [9]	?	?	5	?	1	15	h_r	FVG
Great Britian	Waxed paper	A	1.5	≈0	1	10(4)	h	none
Netherlands	Waxed paper	A	1.5	≈0	1	10	h	none
Romania[8] [9]	Tapered plexiglass	B	-13	in situ	1	15	h_r	CBR_T
Switzerland I[10]	Steel CBR mold	A	-4	≈0	1	≈1	--	CBR_T
Switzerland II[11]	Cellulose foil	A	-4	≈0	1	≈3	h_r	--
CRREL I[12]	Tapered plexiglass	B	variable	0.5	1	12	h_r	none
CRREL II[5]	Plexiglass rings/ rubber membrane	A	26.5	0.5	2	5	h	CBR_T
USA National Crushed Stone Association[13]	Polyethylene film	A	10.5	0.2	1	≈8	h_r	none
USA New Hampshire[14]	Plexiglass rings	A	25	0.5	1	0.5	h_r	none

Notes: [1] A = constant cold plate temperature; B = constant rate of frost penetration.
[2] °F = 9/5°C + 32
[3] 1.0 psi = 6.9 kPa
[4] h = total frost heave; h_r = heave rate; h_p = heave pressure;
     p = ratio of frost heave to square root of freezing index; CBR_T = CBR after thawing
[5] New test being evaluated
[6] Uses separate Hveem R-value test on field samples
[7] Recent changes may have been made
[8] Few details known
[9] Extent of use uncertain
[10] For thaw weakening only
[11] For frost heave only
[12] Currently specified
[13] Abandoned use
[14] Limited use

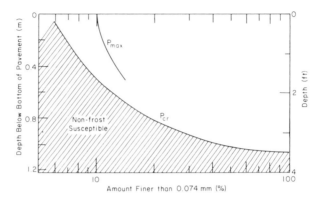

Figure 2.   State of Alaska frost susceptibility criteria.
(P_{cr} is the critical fines content and P_{max} the maximum
allowable fines).

brief summary of the frost susceptibility compliance requirements in
each state follows and is given in Table 2.

   Alaska. The Alaska Department of Transportation and Public Facilities (13) has recently adopted a frost susceptibility criterion that
limits the percent finer than 0.074 mm as a function of depth beneath
the pavement surface (Fig. 2). This criterion assumes a 5 cm thick asphalt concrete pavement on unstabilized gravel base and subbase layers
with essentially non-plastic fines (PI < 6%). The line marked P_{cr}
(critical fines content) is used to restrict excessive pavement deflection during and after thawing. The P_{cr} allowable increases from 6%
finer than 0.074 mm in the upper 3 in. (7.5 cm) of the pavement section
to 100% at 40 in. (1 m) depth. Materials with fine contents falling to
the left of this line are considered non-frost-susceptible. The percent
finer than 0.074 mm can exceed P_{cr} but not P_{max} in the upper 20 in.
(0.5 meters). When the amount of fines exceed P_{cr}, the pavement
thickness is increased if the design traffic considerations warrant.
This criterion was developed by correlating the amount of soil fines
with pavement performance factors, including fatigue cracking, rutting,
and maximum springtime Benkelman beam deflection level. This criterion
is intended to mitigate the effects of reduced subgrade strength associated with freeze-thaw cycling. Surface roughness due to frost heave may
not be eliminated.

   Esch et al. (12) have also reported that Alaska has investigated
the reliability of both frost heave and Hveem R-value strength tests.
The frost heave test employed a fixed cold-plate temperature of 14°F
(-10°C). Segmented ring molds were used to confine the samples and to
minimize side friction. Frost suceptibility was determined on the basis
of the heave rate occurring between 48 and 72 hours after the start of
freezing. More details are given in Table 3. For evaluation of thaw
weakening, the R-value test was used on field samples, and results were
compared to actual pavement performance.

In evaluating the field performance of these tests, they found the best performance to be related to heave rates less than 0.1 in./day (3 mm/day) for base and subbase materials. However, because of the large variation in the results, they concluded that there was insufficient evidence for adopting this test as superior to particle size criteria. Furthermore, they found that the R-value test does not differentiate between materials of low and high frost susceptiblity at all. Thus, they have not recommended the adoption of either of these tests for regular use. However, they have adopted the freezing test for evaluating the use of stabilizing agents.

Arizona. The Arizona Department of Transportation uses the U.S. Army Corps of Engineers (34) criteria for frost susceptibility and pavement design. Good underdrainage is required. Frost heave is not a significant problem in Arizona roads. However, they reported that the Corps of Engineers criteria exclude many materials otherwise acceptable.

Connecticut. The Connecticut DOT uses a modified form of the Casagrande (4) criteria with the special restrictions that less than 10% can be finer than 0.074 mm and that the fines must be non-plastic. No recent details are available.

Illinois. The Illinois DOT uses the U.S. Army Corps of Engineers (34) frost susceptibility criteria for subgrade materials. Base and subbase materials are limited to those with which the past experience has been good. Depth of the water table and the depth of freezing are also taken into consideration. They experience varying degrees of thaw weakening, frost heave, and differential frost heave throughout the State. Regional factors play a large part in their assessment of potential frost damage, as Illinois geography is such that climatological effects range from those of lower Wisconsin to those of the mid-south. Thus, they need to apply a great deal of empirical reasoning to the identification of frost susceptible soils. The Illinois DOT reports that frost heave remains a problem particularly in older secondary roads because of inadequate funding for rehabilitation.

Indiana. The Indiana Department of Highways does not routinely perform frost susceptibility tests. However, for subbase materials the percent finer than 0.074 mm is limited to 8% or no more than 2/3 of the percent finer than 0.59 mm. If problems with frost heave are anticipated, this criterion is augmented by the U.S. Army Corps of Engineers (34) frost susceptibility criteria. However, frost heave is not considered to be responsible for a significant number of performance problems in roads in Indiana.

Iowa. The Iowa DOT uses a complex frost susceptibility criterion based on soil properties. All subgrade materials with >3% organic carbon are considered frost susceptible, as are 1) A-1 through A-4 (AASHTO soil classification system) materials occurring as pocket or interbedded layers bounded by less permeable soils, 2) A-4 soils having a plasticity index (PI) less than 8%, 3) A-2 soils with greater than 15% silt and clay and a PI greater than 4%, and 4) A-7-6 (19 and 20) soils. For use as base and subbase materials, crushed stone is limited to less than 15% finer than 0.074 mm and a PI <6%, and gravels are limited to less than 15% finer than 0.074 mm and a PI <4%. A-6/A-7-6 soils with less than

45% silt and an AASHTO T-99 Proctor density greater than 110 lb/ft^3
(1.76 Mg/m^3) are sometimes used as subgrade replacement materials. The
Iowa DOT reports that this system is successful and no longer considers
frost heave as a major problem.

    Maine. The Maine DOT has developed frost susceptibility criteria
based on the Corps of Engineers (34) method. Grain size curve envelopes
shown in Figure 3 have been developed for use in determining the frost
susceptibility of all soil and granular materials. The percent finer
than 0.02 mm is used as the critical grain size factor, all soils with
less than 3% finer than 0.02 mm being classified as non-frost-susceptible. Materials are assigned frost susceptibility classes of 0, I, II,
III and IV in order of increasing frost susceptibility according to
Table 4 and Figure 3. This method has been in use since the 1950's.

    Massachusetts. The Massachusetts Department of Public Works
replaces subgrade soils that have more than 12% finer than 0.074 mm with
a special granular borrow having less than 10% finer than 0.074 mm to
minimize frost heaving. Granular base and subbase materials can have no
more than 8% finer than 0.074 mm. The Massachusetts DPW has been
actively seeking a more definitive test and as a result has funded
several laboratory and field studies in recent years.

    Michigan. Routinely, the Michigan DOT uses a visual pedological
soil classification system to aide district soils engineers in locating
frost susceptible subgrade soils. Granular base materials are limited
to no more than 8% finer than 0.074 mm and subbase materials to no more
than 7% finer than 0.074 mm, except in certain areas of the state, where
up to 10% finer than 0.074 mm is allowed. For research and experimental

Table 4.   State of Maine Department of Transportation Frost Susceptibility Crite

Frost Class	Kind of soil	Amount finer than 0.02 mm (% by weight)	Frost Susceptibility
0	a) Gravels	0-3	Non-frost-susceptible
	b) Sands	0-3	
I	Gravels	3-10	
II	a) Gravels	10-20	
	b) Sands	3-15	
III	a) Gravels	>20	
	b) Sands, except fine silty sands	>15	Increasing frost susceptibility
	c) Homogeneous clays	PI >12%	
IV	a) Sands, sandy silts, & fine silty sands	>15	
	b) Varved clays and lean clays	PI <12%	

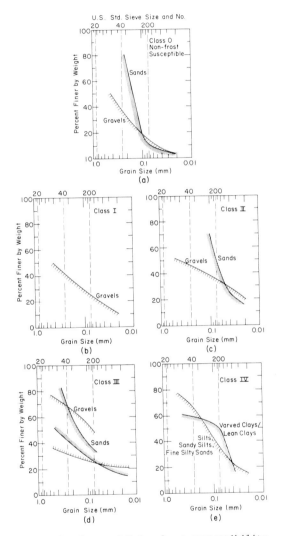

Figure 3. State of Maine frost susceptibility criteria (frost classifications are for shaded side of curves; more details are given in Table 4).

field projects special drainage tests are conducted. Base and subbase
materials are assumed to heave up to 10 percent of their thickness if
the drainage tests indicate the material to be more than 90 percent
saturated when drained at 3 in. (750 Pa) of moisture tension. Frost
heave generally does not occur in the base and subbase layers if the
materials are less than 90 percent saturated at 3 in. (750 Pa) of moisture tension, and the degree of drainage is 50% or greater within 10
days. A laboratory freezing test, based on the Corps of Engineers
method (24), is used to determine the cause of premature pavement distress, but is not used for construction control because of the lengthy
testing time required. Base and subbase gradation requirements accept
materials that, when placed at 4 and 10 in. (10 and 25 cm) thicknesses,
respectively, may produce frost heaves as much as 1 in. (2.5 cm). Little, if any, frost heave has been observed to occur as a result of subgrade freezing when these design practices have been employed.

Minnesota. The Minnesota DOT uses a textural classification system
to identify frost susceptible subgrade soils. Essentially, all fine
grained soils, especially silts, are considered frost susceptible. The
Department is more concerned about thaw weakening and differential frost
heave than the absolute magnitude of frost heave. Thus, base and subbase materials are seldom constructed to the full depth of frost penetration but frost susceptible subgrade soils are mixed to the depth of
frost to minimize the potential for differential heave. Wet areas are
typically subcut and backfilled to the depth of frost with granular
material containing less than 7-15% finer than 0.074 mm. In such situations, perforated pipe is also installed. Base aggregates are limited
to material containing less than 7-10% finer than 0.074 mm. The above
practices have generally proven satisfactory.

Montana. The Montana Department of Highways has adopted the
Casagrande (4) criteria for subgrade materials. All uniform soils with
greater than 10% finer than 0.02 mm and all well-graded soils with
greater than 3% finer than 0.02 mm are considered frost susceptible.
Base and subbase materials have standard specifications that limit frost
susceptibility. Design practice requires roadway grades high above the
water table. No details are available.

New Hampshire. The New Hampshire Department of Public Works and
Highways considers all fine-grained subgrade soil to be frost susceptible. If sands or gravels are encountered the amount finer than 0.074 mm
(of the portion finer than 0.42 mm) is limited to 12 percent. The
restriction also applies to base and subbase materials.

The Casagrande (4) criteria are preferred to those being used. However, because the time required to conduct the hydrometer and specific
gravity tests necessary to determine the percent finer than 0.02 mm is
considered to be too long, the Casagrande criteria are not used. The
standard practice is to employ their 0.074 mm criteria and experience.

The New Hampshire DPW does not appear to have much confidence in
their frost susceptibility criteria. As a result they have sponsored
the development of a rapid frost heave test (36). The essential features of this test are given in Table 3. It employs a top down freezing
mode with a fixed cold plate temperature of 25°F (-4°C). Multiple rings

Table 5. New Hampshire (36) Freezing Test Frost
Susceptibility Criteria.

Frost Susceptibility	Average Rate of Heave (mm/day)
Negligible	<6.5
Very low	6.5-8.0
Low	8.0-10.3
Medium	10.3-13.0
High	13.0-15.0
Very High	>15.0

Note: Customary units; 1 in./day = 25 mm/day.

are used to minimize side friction. The frost susceptibility criteria
adopted are given in Table 5.

New Hampshire has not adopted this test for routine use. They
report that more research is needed to correlate the test results with
roadway performance.

New York. The New York DOT considers all fine-grained subgrade
material to be frost susceptible. Base and subbase material with
greater than 10% finer than 0.074 mm are also considered frost susceptible. The New York DOT design practice concentrates on providing good
drainage. Their criteria are used primarily to ensure adequate thawed
strength. They report that their experience is especially good with
reconstruction projects.

North Dakota. The North Dakota State Highway Department limits the
percent silt (0.074-0.005 mm) for subgrade materials. Allowable amounts
were not provided in the survey. Base and subbase materials are limited
to 15% finer than 0.074 mm (of -19.1 mm material). Frost heave is not a
major problem in North Dakota. Experience with this method has been
good.

Ohio. The Ohio DOT classifies subgrade soils with more than 50% of
soil particles finer than 0.074 mm and a PI less than 10% as frost susceptible. Frost susceptible silty soils are replaced with granular
materials to a depth of three feet. Standard specifications for granular materials eliminate frost susceptible base and subbase materials.
No details were provided. The Ohio DOT reports that no problems have
occurred since these criteria were adopted.

Oregon. The Oregon DOT classifies subgrade soils with greater than
8% finer than 0.074 mm as frost susceptible. Subbase materials with
greater than 8% finer than 0.074 mm and sand equivalents less than 25%
are also classified as frost susceptible, as are base materials with
sand equivalents less than 30%, liquid limits greater than 33% and plasticity indexes greater than 6%. When the amount finer than 0.074 mm
exceeds 8% in the subgrade, the total thickness of the pavement section
must be equal to at least one-half the depth of frost penetration. The

Oregon DOT reports that in conjunction with this special pavement frost
design requirement, these criteria have worked very well.

Pennsylvania. The Pennsylvania DOT uses the U.S. Army Corps of
Engineers frost design method (34). Frost heaving is still a problem
where their current subbase F2 gradation is used. The Pennsylvania DOT
suggests that the sizes of the four frost groups in the Corps of Engineers design method are too large and that six to eight frost groups
would be better. Pennsylvania has been actively supporting research for
a new test. Laboratory methods employing frost heave stress and moisture tension-hydraulic conductivity tests have been evaluated. The results have been disappointing to date. In the future they would prefer
a frost heave test.

Rhode Island. The Rhode Island DOT has the strictest frost susceptibility classification system in the U.S. Their method is based on
Casagrande's (4) but limits non-frost-susceptible soils to no more than
1% finer than 0.02 mm. Considerable frost heave is expected if this
amount is over 3% for non-uniform soils and over 10% for very uniform
soils. This practice has served Rhode Island adequately, as they report
that frost heave has not been a major problem since its implementation.

Vermont. The Vermont Agency of Transportation does not have a specific procedure for determining the frost susceptibility of soil. However, standard base and subbase materials have specifications that limit
frost suseptibility. Gravels are limited to no more than 8% finer than
0.074 mm (of sand portion), crushed stone to no more than 15% finer than
4.76 mm (of total sample), crushed gravel to no more than 12% finer than
0.074 mm (of sand portion), and dense graded crushed stone to no more
than 10% finer than 0.074 mm (of total sample).

Washington. The State of Washington DOT reported that subgrade
soils with greater than 10% finer than 0.074 mm are classified as frost
susceptible. Standard base materials are considered "frost free." The
amount of particles finer than 0.074 mm is limited to 10% (maximum particle size cannot exceed 2/3 of the layer thickness), the ratio of the
percent finer than 0.074 mm to the percent finer than 0.42 mm is limited
to 2/3 and the sand equivalent to less than 30-35%. The Washington DOT
reports that in concert with their special pavement frost design practice (pavement system is equal in thickness to 1/2 the depth of frost
penetration), this method has eliminated frost problems in their highways.

West Virginia. The West Virginia Department of Highways uses the
Corps of Engineers (34) frost susceptibility criteria and pavement frost
design method for flexible pavements, which allows for reduced subgrade
strength during thawing. No special frost suseptibility criteria are
needed for rigid pavements. They appear to have few problems because of
the heavy pavement thicknesses required.

Wisconsin. The State of Wisconsin DOT has assigned each pedological soil class a frost susceptibility rating based on experience. Subgrade materials with more than 2% silt and clay are considered potentially frost susceptible. Granular subbase materials are limited to
8-15% finer than 0.074 mm (of -4.8 mm material). Crushed gravel aggre-

gates are restricted to 3-10% finer than 0.07 mm (of -25.4 mm material
in the upper layers of base courses and of -38.1 mm material in the lower layers). Crushed stone aggregates are limited to 3-12% finer than
0.074 mm (of -25.4 mm material) in the upper layers of base courses and
to 2-12% finer than 0.074 mm (of -38.1 mm material) in the lower layers. Actual frost susceptibility determinations are also dependent on
the depth to the water table and local experience. This method is reported to be reasonably good in predicting frost problems; however,
time, design and cost limitations have sometimes resulted in less than
ideal solutions.

United States - Other Agencies and Individuals

Other agencies and individuals in the United States also have adopted or proposed frost susceptibility classification methods. The most
widely known of these methods are the Casagrande (4) and U.S. Army Corps
of Engineers (34) frost susceptibility classification criteria. These
methods are listed in Table 2 and summarized below.

Asphalt Institute of North America. The Asphalt Institute reports
that 7% finer than 0.074 mm is their dividing point between non-frost-
susceptible and frost susceptible granular base and subbase materials.

Casagrande. The late Arthur Casagrande (4), while studying the
frost heave problem at the Massachusetts Institute of Technology, found
that non-uniform soils containing more than 3% of grains finer than 0.02
mm, and uniform soils containing more than 10% smaller than 0.02 mm are
frost susceptible. This very early observation is the root of many
frost susceptibility criteria, including those of eight states and the
U.S. Army Corps of Engineers. The Casagrande criteria seldom lead to
adverse experiences, which may explain their popularity. However, they
are very conservative criteria and may eliminate many non-frost-susceptible materials from use in road construction.

National Crushed Stone Association. The National Crushed Stone
Association has been considering a frost heave test to identify frost
susceptible crushed aggregates. Kalcheff and Nichols (23) reported that
this test employed a top-down freezing mode with a fixed cold plate temperature of 10.5°F (-12°C). Polyethylene film was used to confine the
sample while minimizing side friction (Kalcheff, pers. comm.). More
details of this test are given in Table 3. This test has never received
official sanction by NCSA and has recently been abandoned from consideration.

United States Army Corps of Engineers. The Corps of Engineers has
been classifying the frost susceptibility of soils since the 1940s.
Their classification system (34) has evolved over the years from the
original work of Casagrande (4). Its present form is shown in Table 6.
Most inorganic materials with 3% or more of grains finer than 0.02 mm in
diameter are classified as frost susceptible for pavement design purposes. Gravels, well-graded sand and silty sands (especially those with
densities near the theoretical maximum density curve), are considered to
be possibly frost susceptible if they contain 1.5-3% finer than 0.02 mm,
and they must be subjected to a standard frost susceptibility test to
evaluate their behavior during freezing. Uniform sandy soils may have

Table 6. U.S. Army Corps of Engineers (34) Frost Design Classification System.

Frost susceptibility	Frost group	Kind of soil	Amount finer than 0.02 mm (% by weight)	Typical soil type under Unified Soil Classification System†
NFS**	None	(a) Gravels	0-1.5	GW, GP
		(b) Sands	0-3	SW, SP
Possibly††	?	(a) Gravels	1.5-3	GW, GP
		(b) Sands	3-10	SW, SP
Very low to high	F1	Gravels	3-10	GW, GP, GW-GM GP-GM
Medium to high	F2	(a) Gravels	10-20	GM, GM-GC, GW-GM GP-GM
Negligible to high		(b) Sands	10-15	SW,SP,SM, SW-SM, SP-SM
Medium to high	F3	(a) Gravels	>20	GM, GC
Low to high		(b) Sands, except very fine silty sands	>15	SM, SC
Very low to very high		(c) Clays, PI>12	--	CL, CH
Low to very high	F4	(a) All silts	--	ML, MH
Very low to high		(b) Very fine silty sands	>15	SM
Low to very high		(c) clays, PI>12	--	CL, CL-ML
Very low to very high		(d) Varved clays and other fine-grained, banded sediments	--	CL and ML; CL, ML and SM; CL, CH, and ML; CL, CH, ML, and SM

* Based on laboratory frost heave tests.
† G = gravel, S = sand, M = silt, C = clay, W = well-graded, P = poorly graded, H = high plasticity, L = low plasticity.
** Non-frost-susceptible
†† Requires laboratory frost heave test to determine frost susceptibility

as much as 10% of their grains finer than 0.02 mm without being frost susceptible.

Soil classified as non-frost-susceptible may heave measurably under field conditions. However, few detrimental effects of frost heaving or thaw weakening would be expected.

The four frost groups (F1, F2, F3 and F4) are used in the Corps of Engineers pavement frost design method. Table 6 and Figure 4 show that there is a considerable range of frost susceptibility within frost

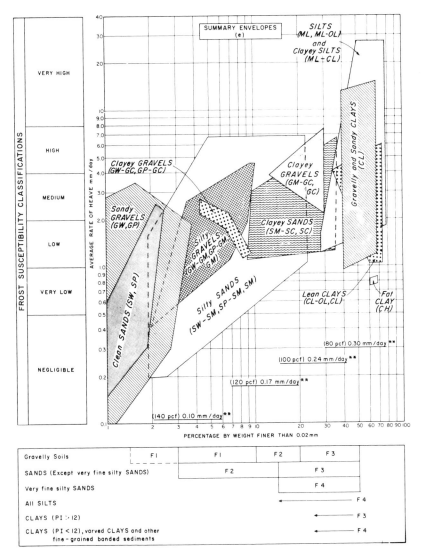

Figure 4.   Degree of frost susceptibility of soils according to the U.S. Army Corps of Engineers (F1, F2, F3 and F4 classifications are used in the Corps pavement design method and do not necessarily indicate the degree of frost susceptibility.

Table 7. U.S. Army Corps of Engineers (CRREL) Freezing
Test Classification System.

Frost Susceptibility	Average Rate of Heave (mm/day)
Negligible	<0.5
Very low	0.5-1.0
Low	1.0-2.0
Medium	2.0-4.0
High	4.0-8.0
Very high	>8.0

Note: Customary units; 25 mm/day = 1 inch/day

groups. Esch et al. (12) found that this system was only weakly related
to pavement performance; they attributed the poor correlation to the
variability in the tabulation. This variability probably reflects the
effects of differences in grain size distribution, dry density, mineral-
ogy, etc., which are not accounted for in the basic frost susceptibility
classification system. This is not necessarily a problem, since the
Corps of Engineers lists all the soil properties and frost classifications in tabular form (24). By comparing the properties of a soil in
question with those of a similar soil in the tabulation, the relative
degree of frost susceptibility can be determined.

A more direct frost susceptibility classification can be made by
conducting a frost neave test. Recent details of the U.S. Army Corps of
Engineers CRREL frost heave test have been given by Chamberlain (8). A
summary is given in Table 3. This test, frequently referred to as the
CRREL frost heave test, uses a variable cold-side temperature to freeze
the sample at a constant rate of frost penetration. A tapered cylinder
is used to minimize side friction. The frost susceptibility classification is given in Table 7.

Because this test is time consuming, requires regular temperature
adjustments, and has difficulties related to side friction, the Corps of
Engineers (6, 7) is conducting research on a new frost heave test. This
new test being developed at CRREL employs a fixed cold-plate temperature
during freezing, and segmented rings to minimize side friction. The
test includes two freeze-thaw cycles to account for changes imparted by
freezing and thawing, and it can be completed in one week. The frost
susceptibility of soils is determined from the rate of heave during the
second freezing and the CBR loss due to freezing and thawing. No new
frost susceptibility criteria have been developed yet.

United States Department of Transportation. The Federal Highway
Administration in the USDOT limits the frost susceptibility of granular
base and subbase material in its standard specifications. Type A materials are limited to 5% finer than 0.074 mm, type B to 6% finer than
0.074 mm (of −5.08 mm material), type C to 8% (of −3.81 mm material),
type D to 10% (of −2.54 mm material), and type E to 11% (of −1.91 mm
material).

Table 8. Summary of Frost Susceptibility Classification Methods in Canada.

Province	Type of Classification	Amount allowable finer than (%) 0.074 mm	Amount allowable finer than (%) 0.02 mm	Allowable plasticity index (%)	Other restrictions
Alberta	I,II	10	3,10	6	—
British Columbia	I,III	8-9	—	—	Frost heave test
Manitoba	II	25	—	12	Grain size distribution
New Brunswick	II	7-30	—	—	Textural & soil class.
Newfoundland	I	6-8	—	—	Type of material
Nova Scotia	II	—	—	—	Textural chart
Ontario	II	10	—	0	% bet. 0.005 & 0.075 mm
Prince Edward Island	I	45	—	—	—
Quebec	I	10	—	—	% finer than 0.053 mm
Saskatchewan	II	10-20	3,10	6	Depth of water table

The FHWA is currently supporting research at CRREL to develop a new freezing test and a model for predicting frost heave.

Canada Provincial Departments of Transportation

All of the Canadian DOTs employ frost susceptibility criteria except the Northwest and Yukon Territories, which are more concerned with the stability of thawing permafrost than with frost heave.

Only three of the Canadian DOTs use exclusively a Type I frost susceptibility classification, while six use a Type II and one a Type III. As in the United States, the preferred critical particle size is 0.074 mm.

A summary of each frost susceptibility classification method is given in Table 8. Details for each province are given below.

Alberta. The Alberta Transportation Department uses the Corps of Engineers (8, 24) summary envelope (Fig. 4) for determining the frost susceptibility of subgrade soils and granular base and subbase materials. They place further restrictions of no more than 10% finer than 0.080 mm and a plasticity index of less than 6% for base courses. Frost heave is not a major problem in Alberta because frost penetrates rapidly and deeply.

British Columbia. The British Columbia Ministry of Transportation and Highways has no frost susceptibility classification method for subgrade soils. Standard specifications generally eliminate frost susceptible granular base and subbase materials. Crushed granular base materials are limited to 9% finer than 0.074 mm (of -37.5 mm material). Select granular subbase materials are restricted to no more than 8% finer than 0.074 mm (of -75 mm material). Crushed granular base materials are limited to 9% finer than 0.074 mm (for -37.5 mm and -75 mm specification materials). Crushed granular surfacing materials for unpaved roads are also restricted to this 9 % limitation; however, the maximum particle size is limited to 19 mm.

A frost heave test is employed on base material when its frost susceptibility is suspect. This test is derived from the British Transport and Road Research Laboratory "LR90" test (10, 30).

Samples are frozen from the top down with a fixed cold side temperature of -17°C. Plastic film is used to contain the sample and minimize side friction. Frost heave is observed for 10 days. The degree of ice lensing and the thaw consistency are observed, but no specific frost susceptibility criteria have been developed.

Manitoba. Subgrade and subbase materials in Manitoba are considered to be potentially frost susceptible if all of the following characteristics are met: 1) 25% or greater of particles are finer than 0.075 mm; 2) the PI is less than or equal to 12%; 3) no more than 25% of particles are finer than 0.075 mm, 4) greater than 60% of particles are between 0.425 mm and 0.005 mm, and 5) less than 20% are coarser than 0.425 mm. These criteria were recently updated to ensure the exclusion of borderline soils that are thought to be potentially frost suscepti-

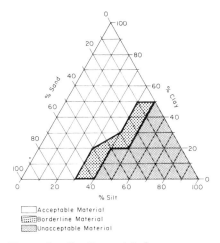

Figure 5.   New Brunswick frost susceptibility criteria.

ble.   There are no frost susceptibility criteria for base materials.
Manitoba reports a great deal of confidence in their procedures.   Problems related to frost heave are reported to be the result of reconnais-
sance, design and construction deficiencies, not the incorrect assessment of the frost susceptibility of materials.

New Brunswick.   The chart in Figure 5 is used by the New Brunswick
DOT to classify the frost susceptibility of subgrade materials.   Subbase
materials are limited to 7% finer than 0.074 mm.   The New Brunswick DOT
reports that severe differential frost heaving has been practically
eliminated on highways using these criteria.

Newfoundland.   The percent finer than 0.075 mm in subgrade soil is
restricted by the Newfoundland DOT.   Experience is apparently the key in
this analysis.   Crushed granular base and subbase materials are limited
to 8% finer than 0.075 mm, while pit gravels are restricted to 6% finer
than 0.075 mm.

Nova Scotia.   The Nova Scotia DOT reports that frost susceptible
subgrade soils are identified using a textural chart and the percentage
of silt and very fine sand.   Standard gradation specifications for granular base and subbase materials eliminate frost susceptibile materials.
No details were given.

Ontario.   The Ontario highway officials use the percentage of particles between 0.005 and 0.075 mm as criteria to determine frost susceptibility of subgrade soils.   The degree of frost susceptibility is given
in Table 9.

Non-frost-susceptible subbase and base materials are limited to no

Table 9. Ontario Frost Susceptibility Criteria for Subgrade Soils.

Percent of Particles Between 0.005 and 0.075 mm	Frost Susceptibility
<40	Low
40-55	Borderline
55-100	High

Table 10. Prince Edward Island Frost Susceptibility Criteria for Subgrade Soils.

Percent of Particles Finer than 0.074 mm	Acceptability
0-45	Acceptable
45-50	Borderline
50-100	Unacceptable

more than 10% finer than 0.075 mm and a plasticity index of zero.
Ontario reports that these criteria have been in use for three years.
Experience with this method is reported to have been satisfactory.

Prince Edward Island. The frost susceptibility of subgrade soils
is classified by the Prince Edward Island Department of Highways and
Public Works as acceptable or unacceptable based on Table 10. Base and
subbase materials are limited to no more than 7% finer than 0.074 mm.
These criteria are usually relaxed on secondary roads because of economic factors. Special attention is given to drainage of granular base
materials and to transition points. Design practice requires the
removal and replacement of frost susceptible subgrade soils only to a
depth sufficient to maintain adequate strength during the spring. Subgrade soils are generally homogeneous, which makes the probability of
differential frost heave low. The current practice appears to be
adequate.

Quebec. No recent details of Quebec's frost susceptibility criteria are available. However, Johnson et al. (20) reported that Quebec
has classified subgrade soils as frost susceptible when more than 10% of
the particles are finer than 0.074 mm and more than 3% are smaller than
0.053 mm.

Saskatchewan. Saskatchewan uses the Casagrande (4) criteria as a
guide to the frost susceptibility of subgrade soils. The depth of the
water table and the thickness of the pavement system are also considered. Crushed and graded base aggregates are limited to less than 10%
finer than 0.074 mm and a PI of less than 6%. Subbase sands are

Table 11. Summary of Frost Susceptibility Classification Methods Used in Europe.

Country	Type of Classification[1]	Finer than (%) 0.074 mm	Finer than (%) 0.02 mm	Allowable Plasticity index	Other Restrictions
Austria	I,II,III	--	3	--	Mineral type, frost heave & thaw CBR tests
Denmark	I,II,[III]	10	--	--	Water table duration, length of freezing test
Federal Democratic of Republic	I,II,III	--	--	--	% finer than 0.063 mm, soil classification, uniformity coefficient, thaw CBR
Finland	II	--	10	--	Grain size curves, capillarity test
France	II,III	--	--	--	Soil class., frost heave test
German Democratic Republic	II	--	--	--	% finer than 0.10 mm, adsorbed water capacity, mineral chemical activity
Great Brittain	II,III	10-40	--	0-20	Saturated water content, drainage condition, frost heave test
Netherlands	I,II,III	--	3	--	% finer than 0.063 mm, sand equivalent, frost heave test
Norway	I,II	--	3	--	% finer than 0.002 mm & 0.20 mm, saturated CBR test
Poland	[II],[III]	18-40	--	--	Grain size curves, overburden pressure, frost heave test
Romania	II,[III]	--	10	0	% finer than 0.10 mm & 0.002 mm, frost heave and thaw CBR tests
Sweden	I,II	16	--	--	Capillarity & shake tests
Switzerland	I,II,III	--	1.5-10.0	x	% finer than 0.02 mm, uniformity coef., soil class., frost heave & thaw CBR tests

Note: [ ] designates method under consideration, but not adopted.

restricted to 20% finer than 0.074 mm and a PI of 6%. Because of a
cold, dry climate and a scarcity of silty soils, frost problems are
reported not a major concern of Saskatchewan highway officials.

Europe

     Methods to determine the frost susceptibility of soils and granular
materials in Europe appear to be in a continuous state of flux, regularly being improved and updated. These methods frequently are more complicated than those commonly used in North America. For example, of the
12 European nations for which recent information is available, 4 use
frost susceptibility criteria with all Type I, II, and III restrictions,
and 5 more use both Types II and III. None use Type I alone. The most
widely used critical particle size in Europe is the 0.02 mm. This is
clearly the result of Casagrande's work and is in sharp contrast to the
0.074-mm size preferred in North America. These test methods are listed
in Table 11 and summarized below.

     Austria. Brandl (3) developed frost susceptibility criteria for
gravels in Austria based on the 0.02 mm grain size and the mineral
type. This classification is given in Table 12. He has also developed

Table 12. Austrian (3) Frost Susceptibility Criteria for Gravel.

Maximum Allowable Percentage of Grains <0.02 mm	Allowable Mineral Composition of Non-Frost-Susceptible Soils
3	Non-frost-susceptible, no mineral type determination necessary

5	Normally, if heave properties are known from field or laboratory observations, no mineral type determination is necessary. If frost heave properties are not known, the gravel is non-frost-susceptible if
	  1) the minerals are inactive or
	  2) there is a mixture of the inactive minerals and a maximum of
	     a) 10% kaolinite
	     b) 30% chlorite
	     c) 30% vermiculite
	     d) 40% montmorillonite, and/or
	     e) 50% mica,
	with boundary conditions of
	     a) 60% mica and chlorite
	     b) 50% mica, chlorite and kaolinite
	     c) 50% mica and kaolinite
	     d) 40% mica, chlorite, kaolinite and montmorillonite.
	In addition, up to 40% complex silicate is allowable.
	  3) If evidence of iron hydroxide, frost heave tests are required.

8	Inactive minerals with 1% $\leq$ 0.002 mm.

a laboratory frost heave test. The test employs a constant cold side temperature of -15°C and freeze-thaw cycling. Multiple stacked rings are used to contain the samples and to minimize side friction. Both frost heave and thaw weakening characteristics are assessed. A range of 0.4 to 0.8 in. (1 to 2 cm) of frost heave is allowed for main roads and 1 inch (2.5 cm) for secondary roads. A minimum thaw CBR of 25% is allowed for main highways, 20% is allowed for secondary roads. More details are given in Table 3.

Denmark. The Danish State Road Laboratory (9) restricts non-frost-susceptible soils to no more than 10% finer than 0.075 mm. They also rank soil groups according to the field conditions of water table depth and length of the freezing season. Few details of this classification system are available. Denmark also has been considering a laboratory frost heave test similar to the British TRRL LR90 test (10, 30). A constant cold-side temperature of 1.5°F (-17°C) is used. The sample stands alone with no support to eliminate side friction. Thaw CBR and frost heave at 240 hours are used to assess frost susceptibility. More details are given in Table 3.

Federal Republic of Germany. The West German frost susceptibility criteria (1) have been regularly improved during the past decade. The present criteria are given in Table 13.

These frost susceptibility criteria are principally based on thaw weakening. If any further questions remain about a soil's frost susceptibility, laboratory freezing tests are specified. The factors observed

Table 13. West German (1) frost susceptibility criteria.

Frost susceptibility	Soil classification* (West German Standards)	Allowable amount (%) finer than 0.063 mm
None	GW, GI, GE, SW, SI, SE	5
	TA	--
	OT, OH	--
Low-medium	TM	--
	ST, GT	5 if C_u =15; 15 if $C_u \leq$ 6†
	SU, GU	5 if C_u =15†; 15 if $C_u \leq$ 6†
High	TL	--
	UL, UM	--
	ST, GT	--
	SU, GU	--

* Listed in order of increasing frost susceptibility; G = gravel, S = sand, U = silt, T = clay, O = organic, H = peat, A = high plasticity, M = medium plasticity, W = well-graded, I = intermediate gradation, E = skip-graded, T = very clayey, U = very silty.
† If 6 < C_u < 15, then the allowable amount finer than 0.063 mm should be linearly interpreted between 5 and 15%.

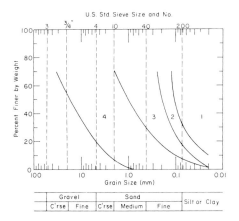

Figure 6. Finnish frost susceptibility
criteria (zone 1, frost susceptible;
zones 2-4, frost susceptible only if
lower part of gradation curve extends
into an adjacent lower numbered zone).

include frost heave, heaving pressure, rate of heave, and loss in bearing capacity after freeze-thaw cycling. Few details of this test are
available.

Finland. The Roads and Waterways Administration in Finland (19)
assesses the frost susceptibility of soils using the gradation chart
shown in Figure 6. All soil types whose gradation curves lie in area 1
are considered frost susceptible. Those soils whose gradation curves
are in areas 2, 3, or 4 are considered non-frost-susceptible if the
lower parts of their curves do not extend into an adjoining finer-
grained area. In borderline cases a capillarity test is conducted. The
capillarity of non-frost-susceptible soils should be below 40 in. (100
cm.)

France. The French Regional Road Research Laboratory of Nancy
reports (26) the frost susceptibility classification method shown in
Table 14. This frost susceptibility classification table is offered

Table 14. French (26) frost-susceptiblity classification method.

Soil Classification*	Frost Susceptibility
D_1, D_2, D_3, D_4	None
A_4, B_1, B_2, B_3, B_4	Low
A_1, A_2, A_3, B_5, B_6, E_1, E_3	High

* According to French soil classification system.

Table 15. Classification of the Frost Susceptibility of Soils
According to the French Frost Heave Test.

$\Delta x_g / \Delta\sqrt{I}$ $\mathrm{mm}/(°\mathrm{Ch})^{1/2}$	Frost Susceptibility
<0.05	None
0.05-0.40	Little
>0.40	High

Note: $\Delta x_g / \Delta\sqrt{I}$ given in original text in SI:
1 $\mathrm{mm}/(°\mathrm{C\cdot h})^{1/2} = 0.053$ in./$(°\mathrm{F\cdot h})^{1/2}$.

instead of the laboratory frost heave test from which it is derived.
The frost heave test, which remains the preferred method for assessing
the frost susceptibility of soil, has been conducted on approximately
500 soils. This test employs top-down freezing with a constant cold
plate temperature of 22°F (-5.7°C). Side friction is minimized by using
a lubricated foam rubber tube as a liner in the cylindrical cell used to
confine the sample. The slope of the frost heave x_g versus square
root of the freezing index I curve is used as the critical frost heave
factor. The frost susceptibility is assessed as shown in Table 15.

German Democratic Republic. Klengel (25) proposed the frost susceptibility classification system for aggregates shown in Table 16 for
use in East Germany. This table was developed from data obtained from
laboratory freezing tests and is based on both frost heave and thaw
weakening. Few details of the frost heave test are available. However,
it is known that their cold-plate temperature is as low as -15°C, that
the test is conducted with both open and closed water supplies, and that
the critical factors for frost susceptibility are the frost heave ratio
and the frost variability level (percent increase in water content).
The last factor is assumed to characterize loss in bearing capacity.
The known details are summarized in Table 3.

Table 16. East German (25) Frost Susceptibility Criteria.

Gravel type	Particle <0.1 mm in diameter (%)	Absorbed water capacity	Mineral chemical activity	Frost heave susceptibility	Bearing capacity reduction during thawing
Coarse-grained aggregate	<10	<0.25	Low to high	None	None
	10-30	<0.30	Low	Variable	Slight
	30-50	>0.30	Low	Slight	Slight to moderate
Fine-grained aggregate	50-75	0.30-0.50	Low	Slight to moderate	Moderate to high
	>75	0.50-0.80	High	Slight to very high	Slight to moderate
		>0.80	High	Slight	Slight

Table 17. British TRRL Freezing Test
Frost Susceptibility Classification
Criteria.

Frost heave at 240 hr, (cm)	Frost Susceptibility
≤ 1.3	None
$1.3-1.8$	Marginal
>1.8	Highly

Table 18. Non-Frost-Susceptible British Road Construction Materials.

Material Type	Amount Finer than 0.075 mm (%)	Plasticity Index (%)	Saturated Water Content (%)
Cohesive soils[1]	--	>15	--
Cohesive soils[2]	--	>20	--
Non-cohesive soils	≤ 10	0	--
Limestone gravel	--	--	≤ 2
Crushed oolitic and magnesian limestone	--	--	≤ 3
Crushed hard limestone	≤ 10	--	--
Crushed chalk	--	--	--
Crushed granites	≤ 10	--	--
Crushed graded slag	--	--	--
Coarse pulverized fuel ash	≤ 40	--	--

[1] Good drainage
[2] Poor drainage, water table within 2 ft (0.6 m) of material.

Great Britain: The British Transport and Road Research Laboratory
(TRRL) (10, 30) has developed a freezing test for assessing the frost
susceptibility of road construction materials. This test has been used
since 1945, which makes it one of the oldest freezing tests still in
use. Materials are frozen with a fixed cold-side temperature of $-17°C$.
Side friction is minimized using waxed paper. The critical frost
susceptibility factor is the amount of frost heave after 240 hours. The
frost susceptibility criteria are given in Table 17.

Considerable effort (20) has been made in recent years to improve
on the test equipment and the reliability of the test results. A recent
modification under consideration (30) is to shorten the freezing period
from 10 to 5 days.

As a result of experience with the TRRL test, certain materials
have been identified as being non-frost-susceptible for the most severe
winter conditions experienced in England. Table 18 lists these soils.
Two other British road construction materials (burnt colliery shale and
fine pulverized fuel ash) require a freezing test to assess their frost
susceptibility.

Netherlands. The Dutch Study Center for Road Construction (28) has been intensively investigating frost susceptibility criteria in the Netherlands during the past few years. The Casagrande (4) criteria are presently being used for fine grained subgrade soils. Sandy subbase soils are frost susceptible if more than 15% of the particles are smaller than 0.063 mm. If 10-15% are finer than 0.063 mm, then no more than 3% can be finer than 0.02 mm.

To improve their reliability in determining frost suscpetibility, the Dutch investigated using the British TRRL freezing test, but concluded that it was too time consuming. As a result, they decided to pursue a faster method and selected the sand equivalent test, which they felt would address the percentage of fines factor common to most frost susceptibility criteria and the minerological character of the clay. They concluded that if the sand equivalent SE value is greater than 30, then the soil is non-frost-susceptible. If the SE value is less than or equal to 30, a freezing test is required.

Norway. The Public Roads Administration (29) requires that base and subbase materials in ordinary bituminous surfaced roads must have less than 3% finer than 0.02 mm to be non-frost-susceptible. Gravel-surfaced roads with an average daily traffic of less than 50 can have as much as 5% finer than 0.02 mm.

The frost susceptibility of subgrade soils in Norway is classified as shown in Table 19. This frost susceptiblity classification is based on loss of bearing capacity during thawing.

The Norwegians (17) are considering a laboratory CBR-water-content test for determining the frost susceptibility of granular materials. If the CBR is not seriously reduced by soaking, then the material is considered non-frost-susceptible. Gaskin (17) reported that the allowable percent finer than 0.02 mm for one gravel could be increased from 3 to 5% as a result of this type of analysis. Recently Geir Refsdal (pers. comm.) of the Norwegian Road Research Laboratory reported that this method is now being used in special cases where there is a shortage of gravel with less than 3% finer than 0.02 mm.

Poland. The frost susceptibility classification method shown in Figure 7 has been proposed (27) for use in Poland. This classification

Table 19. Frost Susceptibility Classification of Subgrade Soils in Norway (29).

Degree of frost susceptibility	% finer than: (of material < 19 mm)		
	0.002 mm	0.02 mm	0.2 mm
None		<3	
Little		3-12	
Moderate	*	>12	<50
High	<40	>12	>50

* Soils with more than 40% < 2 μm are moderately frost susceptible.

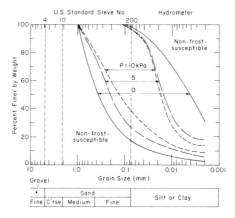

Figure 7.   Example of proposed frost susceptibility criteria for Poland (limits
vary with freezing temperature).

is the result of laboratory freezing tests. It is similar in concept to
the Alaskan criteria in that the grain size limits are modified by the
depth or amount of surcharge. Figure 7 is for a mean freezing temperature of 23°F (-5°C); at lower freezing temperatures the frost susceptible zones become smaller.

Romania. The Romanian frost susceptibility standards (35) are
based on grain size and Atterberg limits. Table 20 shows the standards
now in use. In addition a laboratory freezing test is being considered

Table 20.   Romanian (34) Frost Susceptibility Criteria.

Frost Susceptibility	Type of Soil	Plasticity (%)	Particle Diameter (mm)	Percentage of the Total Specimen Mass
			Criteria	
			Grading	
None	Non-cohesive soil without clay	PI = 0	<0.002 <0.02 <0.1	<1 <10 <20
Low-high	Non-cohesive soil with clay	PI < 10 PI > 35	<0.002 <0.02 <0.1	<6 <20 <40
Very high	Cohesive soil	10<PI<35	<0.002 <0.02 <0.1	>6 >20 >40

as a more direct method of assessing frost susceptibility. This test employs a variable cold plate temperature ranging from 32 to -13°F (0 to -25°C) to maintain a constant frost penetration rate of 1 cm per day. Side friction is minimized by using a tapered cylinder. Details are given in Table 3. This test is similar to the CRREL test. The CRREL average-rate-of-heave criteria are suggested for classifying the frost susceptibility of soils. In addition the reduction in CBR due to freezing and thawing is suggested; however, no criteria are given.

Sweden. The Swedish National Road Administration (32) uses a frost susceptibility classification method that relies on the percent finer than 0.075 mm and the capillarity of the soil and a shake test. The capillarity is assessed by determining the vacuum required to cause air to break through a column of saturated soil. The shake test assesses the flow and moisture retention characteristics of a lump of saturated fines. Both the capillarity and shake tests are unique to the Swedish frost susceptibility classification method. The Swedish frost suscepti- bility classification criteria are shown in Table 21. Thoren (31) reported that this method has been used for a long time and that there has been good correlation between field and laboratory results.

Switzerland. The Swiss Association of Road Professionals (14) employs all three frost susceptibility classification levels. The Type I criteria are based on the percent finer than 0.02 mm and are given in Table 21. The Type II criteria are based on the soil type and plastic- ity index as well as the percent finer than 0.02 mm and are given in

Table 21. Swedish (32) Frost Susceptibility Criteria.

Material Type	Amount Finer[1] than 0.075 mm (%)	Capillarity[2] (m)	Shake[3] Test	Frost Susceptibility
All mineral soils	<16	<1.0	--	None
Tills, clay-free % clayey	16-43	--	--	
Till, clayey	16-43	--	negative	
Boulder clays	>16	--	--	Moderate
Sedimentary soils	>16	1.0-1.5	--	
Clays[4]	>16	--	--	
Tills, clay-free, slightly clayey & clayey	>43	--	--	Very high
Tills, clayey	>16	--	positive	
Sedimentary soils	>16	>1.5	positive	

[1] Of portion finer than 16 mm.
[2] Air entry moisture tension value when subject to a vacuum.
[3] If a lump of saturated soil becomes shiny and flows when shaken, but surface returns to dull and firm state when squeezed, then it is frost susceptible and the test is positive.
[4] Can be rolled into a thread less than 1 mm in diameter.

Table 22. First level (Type I) of the
Swiss (14) frost susceptibility criteria.

Frost Susceptibility	Amount Finer Than 0.02 mm* (%)
None†	<1.5
Borderline	1.5-3
High	>3

* Applied only to the fraction smaller than 60 mm.
† Homogeneous sands with $C_u > 5$ are practically non-frost-susceptible if they contain less than 10% finer than 0.02 mm.

Table 22. This criterion has been developed from the US Army Corps of Engineers (34) criteria and remains very similar.

The Type I and Type II criteria are used for all subgrade, subbase and base materials. The Type III criteria are used only for the slightly frost-susceptible gravels listed in Table 23. The Type III criteria

Table 23. Second Level (Type II) Swiss Frost Susceptibility Criteria.

Frost susceptibility	Soil type	Amount (%) Finer Than 0.02 mm	Unified Soil Classification*
Slight	Gravel	3-10	GW, GP GM, GC
Slight to moderate	a) Gravel	10-20	GM, GC-CL GM-GC, GM-ML
	b) Sand	3-15	SW,SP, SM, SC
Moderate	a) Gravel	>20	GC-CL, GM-GC, GM-ML
	b) Sand (except very fine silty sand)	>15	SC-CL, SM-SC, SM-ML
	c) Clays, PI > 12		CL, CH
High	a) Silt		ML, MH
	b) Very fine silty sand	>15	SM-ML
	c) Clayey silt, PI < 12		CL, CL-ML
	d) Banded clays and other banded fine soils		In alternate layers CL, ML, CL, ML, SM, CL, CH, ML, CL, CH, ML, SM

* G = gravel, S = sand, M = silt, C = clay, W = well-graded, P = poorly graded, H = high plasticity, L = low plasticity.

require two separate freezing tests, one to determine the CBR loss after
freezing and thawing and the other to determine the amount of frost
heave occurring in 72 hours. The CBR test sample (15) is frozen in a
CBR mold at a constant cold-plate temperature of $-4°F$ ($-20°C$). No provision is made for reducing side friction. The frost heave test (14) is
also conducted at a constant cold plate temperature of $-4°F$ ($-20°C$).
However, cellulose foil is used to confine the sample and to minimize
side friction. If the CBR after freezing and thawing is greater than
the corresponding AASHTO requirement and the frost heave is not more
than 40 mm, then the gravel is non-frost-susceptible. Based on experience since this standard was adopted, the Swiss criteria are reported
(14) to be conservative, particularly the frost heave test.

THE MOST PROMISING METHODS

    This review has revealed that there are many widely diverse stan-
dards for determining the frost susceptibility of soils and granular
materials. For instance, the tolerances for certain particle sizes,
especially, the 0.074 mm size, are wide ranging; Prince Edward Island
allows as much as 45% finer than 0.074 mm, while Indiana allows no more
than 8%. It is difficult to determine the reasons for these differences. Certainly some of the differences result solely because of the
variety of maximum particle sizes specified for the grain size analyses. We do not have enough information to normalize all of the grain
size data to a certain maximum particle size. If we did, we would probably find that much of the scatter would remain; only the order would be
rearranged. Transcending this problem, we must conclude that the differences in particle size criteria and in frost susceptibility criteria
generally result from the independent way in which most of the procedures were developed and/or evolved. A common ground has not been
obtained because 1) criteria have been developed for regional problems
not common to all jurisdictions, 2) most of the criteria have not been
rigorously validated, and thus, their use cannot be convincingly argued
for adoption by other agencies, and 3) there is a general lack of communication about frost susceptibility criteria (with valid arguments and
willing listeners).

    There also appears to be a diversity in the purposes of the frost
susceptibility standards reported. A large number of agencies reported
that their criteria were used to minimize the effects of thaw weakening,
while others reported that frost heave or differential frost heave was
their problem. Still others reported criteria designed simply to minimize frost susceptibility, without specifying what frost susceptibility
problem they were concerned with.

    Another factor affecting the analysis of the frost susceptibility
criteria is the pavement design practice used. Johnson et al. (21)
reported many widely diverse pavement design methods. Since the pavement design practice can compensate for poor frost susceptibility
criteria (and for poor materials), the reliability of the various frost
susceptibility criteria is difficult to judge.

    This lack of common ground makes it difficult to determine which
frost susceptibility method is the most effective without making a

direct comparison. This would require an extensive laboratory and field
program. Moreover, a single method that would reliably predict frost
susceptibility for a variety of conditions would not necessarily be in
the best interest of many road designing agencies. This is because most
design agencies (particularly state and provincial agencies) do not have
problems for a wide range of soil and climatic conditions, but have
narrowly focused problems with particular material types and/or moisture
and temperature conditions. For these agencies, the specialized frost
susceptibility criteria that work best will be the appropriate solution.

Some general conclusions about the merits of the various frost
susceptibility criteria, however, can be drawn. One is that any Type I
frost susceptibility criterion based on a single particle size must be
conservative in order to be successful in screening out frost susceptible materials. The reason for this is that a single particle size cannot directly assess frost susceptibility because the other controlling
factors such as mineral type, density or porosity, pore size distribution, water availability, etc., cannot be addressed solely by particle
size. Previous studies by Chamberlain (6, 7), Gaskin (17, 18), and
Townsend and Csathy (33) have drawn the same conclusion. Nonetheless,
we feel that it is essential that the option of using a Type I criterion
should be available. The most frequently used Type I critical particle
size is 0.074 mm. It is believed that this size has been selected most
frequently because of the ease with which the test can be conducted.
Esch et al. (13) observed a slightly better correlation of pavement performance in the field with the percentage of particles finer than 0.074
mm than with the percentage finer than 0.02 mm, in spite of the fact
that the 0.02 mm particle size was a stronger predictor of frost heave
in the laboratory. This apparent anomoly may be because soils with
clayey fines were rarely encountered in their study; thus, the effects
of plastic fines are unknown. To ensure that the clay component of
materials is addressed the 0.02 mm particle size remains preferable.
Therefore, the Casagrande (4) criteria appear to be the most appropriate. Its reliability in predicting frost susceptibility of soils was
high in two studies (6 and 32). However, Townsend and Csathy have shown
that the Casagrande (4) criteria (and all of the other particle size
criteria evaluated) are not very reliable in predicting non-frost-
susceptible soils; i.e., non-frost-susceptible soils are frequently
classified as frost susceptible. As a result, non-frost-susceptible
soils are frequently rejected for use in road construction when they are
suitable for use.

This problem leads us to the next conclusion: a Type II frost
susceptibility criterion is required to better discriminate frost
susceptible materials. Chamberlain (6) found that this type of particle
size criterion is the most reliable, the Swiss and the U.S. Army Corps
of Engineers methods heading the list. The root of both of these
criteria is the Type I Casagrande criteria. The addition of soil
classification and Atterberg limit data and the direct correlation of
these material characteristics with field and laboratory frost heave and
thaw weakening observations may provide a more reliable determination of
frost susceptibility.

There are certain road building materials, however, for which a reliable characterization of frost susceptibility cannot always be made without a Type III freezing test. These are the granular base and subbase materials containing small quantities of silt and clay. It is for this reason that a freezing test is needed. Because both frost heave and thaw weakening can be major problems with these materials, both need to be addressed in the freezing test. Several of the European agencies require a CBR test after thawing to characterize thaw weakening. However, the only United States test to address thaw weakening directly is the freezing test now being developed at CRREL (6, 7).

The ideal frost susceptibility classification system, then, should include all three types of frost susceptibility classification methods, and the freezing test should include both frost heave and thaw weakening factors. This would allow the frost susceptibility to be determined at the level of effort commensurate with project requirements and acceptable risks. The frost susceptibility classification systems that include elements of all three frost susceptibility classification types include the Austrian, Danish, West German, Swiss and U.S. Army Corps of Engineers methods.

An even greater reliability might be obtained in the Type III test if a laboratory freezing test were conducted under simulated field conditions of surcharge, moisture availability, freezing rate, etc. This might allow predictions of actual amounts of frost heave, rather than the relative degree of frost susceptibility that most freezing tests provide. If the thaw weakening is characterized with a resilient modulus factor as done by Johnson et al. (22), then more precise mechanistic methods might be used to design pavement systems for frost action. One more level in the hierarchy of frost susceptibility classification methods, Type IV, would then be available.

This leads to one last but important consideration in applying frost susceptibility criteria - the effect of the natural variation of in situ conditions of frost heave and thaw weakening. This will become especially important once more accurate predictions of frost heave and thaw weakening can be made. It is at present a weak link in the design of pavement systems for frost action. It is the variability in material properties and moisture conditions that often leads to pavement roughness due to freezing and thawing. Recent studies (5) have devoted considerable attention to this problem, but an application to predicting frost susceptibility has not yet been made.

SUMMARY AND CONCLUSIONS

Frost susceptibility determinations are generally made at three levels of sophistication: Type I, which is primarily based on the percent finer than a specified particle size; Type II, which is based on soil type or classification and sometimes augmented by other information such as the Atterberg limits, capillarity and permeability; and Type III, which requires a laboratory freezing test and observations of frost heave and/or thaw weakening.

The variety of frost susceptibility determination methods surveyed
is nearly equal to the number of agencies employing them. Nearly half
of the transportation agencies in the United States have adopted specific criteria for determining the frost susceptibility of soils and granular materials and nearly all of these are unique. The single most
common thread is the Casagrande (4) criteria. A small majority of the
methods are of the Type II criteria and there is a strong preference for
Type I criteria. None of the state DOTs regularly employs a Type III
frost heave test; the only U.S. agency that does is the U.S. Army Corps
of Engineers.

In Canada the most common critical particle size factor is 0.074
mm, as it is in the United States. A significant majority of the
methods are Type II. One Canadian transportation agency employs a Type
III freezing test.

In Europe the preferred critical particle size is 0.020 mm, and
many of the national transportation agencies employ Type III freezing
tests. Aside from the U.S. Army Corps of Engineers and the State of
Alaska Department of Transportation and Public Facilities, the Europeans
appear to be the most active in developing comprehensive frost susceptibility criteria including all three types of classification methods.
This may be the result of a greater scarcity of non-frost-susceptible
road construction materials in Europe than in the U.S. and the resulting
need to discriminate better in their selection.

As this need is also becoming more evident in the United States, it
is essential that we also adopt more comprehensive frost susceptibility
criteria. These criteria should include all three types to allow the
selection of a method appropriate to the task. Furthermore, it is also
desirable to conduct the freezing test simulating actual field conditions, and to quantify thaw weakening with the resilient modulus, so
that more mechanistic methods of designing pavement systems for frost
action can be used. The criteria selected can be narrowly focused on
regional problems of state or provincial agencies or more widely based
on generalized problems of national agencies. Whatever criteria are
selected, their performance should be well documented and validated, and
revisions should be made when necessary.

ACKNOWLEDGMENTS

Much of review in this paper was performed at CRREL. We wish to
acknowledge the support there given by the U.S. Army Corps of Engineers,
the Federal Highway Administration, and the Federal Aviation Administration. In addition, we wish to thank the State of Alaska, Queen's
University and the Norwegian Road Research Laboratory for their support
of our research into the frost susceptibility of soils. We are also
grateful to the many transportation agencies in the United States and
Canada for co-operating with this survey. Finally we want to acknowledge the thorough editing of the manuscript provided by David Cate of
CRREL.

APPENDIX - REFERENCES

1. Behr, H., "Criteria for the Determination of the Frost-Susceptiblity
    of Soils in the Federal Republic of Germany," Frost i Jord, no. 22,
    Nov., 1981, p. 27-34.

2. Beskow, G., "Soil Freezing and Frost Heaving with Special Application
    to Roads and Railroads," The Swedish Geological Society, Series C,
    No. 375, 26th Yearbook, No. 3, 1935, 145 p. (translated by J.O.
    Osberberg; Published by Technical Institute, Northwestern University,
    Nov., 1974).

3. Brandl, H., "The Influence of Mineral Composition on Frost Suscepti-
    bility of Soils," in Proceedings, Second International Symposium on
    Ground Freezing, Norwegian Institute of Technology, June 24-26, 1980,
    Trondheim, p. 815-823.

4. Casagrande, A., "Discussion of Frost Heaving," Highway Research Board,
    Proceedings, vol. 11, 1931, p. 163-172.

5. Chamberlain, E.J., Guymon, G.L., and Berg, R.L., "Probabilistic
    Approach to Determining Frost Heave," Proceedings, Seminar on the
    Prediction of Frost Heave, Department of Civil Engineering, University
    of Nottingham, April 9, 1981, p. 29-50.

6. Chamberlain, E.J., "Comparative Evaluation of Frost-Susceptibility
    Tests," in Frost Action and Risk Assessment in Soil Mechanics, Trans-
    portation Research Record 809, 1981, p. 42-52.

7. Chamberlain, E.J., " Frost Susceptibility of Soil--Review of Index
    Tests." U.S. Army Cold Regions Research and Engineering Laboratory,
    Monograph 81-2, Dec., 1981, 121 p.

8. Chamberlain, E.J., and Carbee, D.L., "The CRREL Frost Heave Test,"
    Frost i Jord, no. 22, Nov., 1981, p. 55-63.

9. Christensen, E. and Palmqvist, K., "Frost Action in Soils. Theories,
    Criteria, Instruments, Results," (in Danish). Tekniske Hojskole.
    Instituttet for Vejbygning, Trafikteknik og Byyplanlaegning. Report
    no. 4, Eksamensarbejde, 1976, 187 p.

10. Croney, D. and Jacobs,J.C., "The Frost Susceptibility of Soils and
    Road Materials," Transportation and Road Research Laboratory,
    Berkshire, England, Report LR90, 1967, 68 p.

11. Ducker, A., "Untersuchungen über die Frostgefährlichkeit nichtbindiger
    Boden," Forschungsarbeiten aus dem Strassenwesen, vol. 17. Berlin:
    Volk und Reich Verlag, 1939.

12. Esch, D., McHattie, R., and Connor, B., "Frost Susceptibility Ratings
    and Pavement Structure Performance," Transportation Research Record
    809, 1981, p. 27-34.

13. Esch, D.C. and McHattie, R.L., "Prediction of Roadway Strength from Soil Properties," State of Alaska Department of Transportation and Public Facilities, February, 1982, 15 p.

14. Fetz, L.B., "Frost Susceptibility of Soils," Frost i Jord, no. 22, Nov., 1981, p. 39-40.

15. Fetz, L.B., "Short-cut Frost Heaving Test for Soils," Frost i Jord, no. 22, Nov., 1981, p. 41-48.

16. Frost i Jord, "Frost-Susceptibility of Soils - Criteria from Several Countries," no. 22, Nov., 1981, 63 p.

17. Gaskin, P., "Water Susceptibility of Gravel," Internal Report No. 994, Norwegian Road Research Laboratory, September, 1981, 31 p.

18. Gaskin, P., "Review of Frost Susceptibility Classifications," Frost i Jord, No. 22, Nov., 1981, p. 3-10.

19. Hailikari, T., "The Frost Susceptibility Test for Public Roads in Finland," Frost i Jord, no. 22, Nov., 1981, p. 11-12.

20. Jones, R.H., "Developments and Applications of Frost Susceptibility Testing. In Proceedings, Second International Symposium on Ground Freezing, Norwegian Institute of Technology, June 24-26, 1980, Trondheim, p. 748-759.

21. Johnson, T.C., Berg, R.L., Carey, K.L. and Kaplar, C.W., " Roadway Design in Seasonal Frost Areas," U.S. Army Cold Regions Research and Engineering Laboratory (CRREL) Technical Report 259, 1975, 104 p.

22. Johnson, T.C., Cole, D.M., and Chamberlain, E.J., "Influence of Freezing and Thawing on the Resilient Properties of a Silt Soil Beneath an Asphalt Concrete Pavement," CRREL Research Report 78-23, Sept. 1978, 53 p.

23. Kalcheff, I.V. and Nichols, F.P., "A Practical Test Procedure for Frost Heave Evaluation of Granular Base and Subbase Materials." Symposium on Graded Aggregate Bases and Base Materials, American Society for Testing and Materials, 1974.

24. Kaplar, C.W., "Freezing Test for Evaluating Relative Frost Suscepti-bility of Various Soils." CRREL Technical Report 250, 1974, 40 p.

25. Klengel, K.J., "Recent Investigations on Frost Heaving," (in German). Wissenschaftliche Zeitschrift der Hochschule fuer Verkehrswesen "Friedrich List," Dresden, vol. 17, no. 2, 1970, p. 372-377. (Also CRREL Draft Translation 383.)

26. Livet, J., "Experimental Method for the Classification of Soils According to Their Frost Susceptibility, France," Frost i Jord, no. 22, Nov., 1981, p. 23-26.

27. Pietrzyk, K., "Attempts of a New Formulation on the Criterion of Ground Freezing," Proceedings, Second International Symposium on Ground Freezing, Norwegian Institute of Technology, June 24-26, 1980, Trondheim, p. 795-806.

28. Post, H.J., "Frost Heave Prediction in the Netherlands," Proc. of seminar, The Prediction of Frost Heave, Dept. of Civil Engineering, University of Nottingham, April 9, 1981, p. 23-28.

29. Saetersdal, R., "Prediction of the Frost Susceptibility of Soils for Public Roads in Norway," Frost i Jord, no. 22, Nov., 1981, p. 35-36.

30. Sherwood, P.T., "British Experience with the Frost-Susceptibility of Roadmaking Materials," Frost i Jord, no. 22, Nov., 1981, p. 49-54.

31. Taber, S., "Frost Heaving," Journal of Geology, vol. 37, vol. 10, no. 3, 1929, p. 553-557.

32. Thoren, H., "Prediction of the Frost Susceptibility of Soils for Public Roads in Sweden," Frost i Jord, no. 22, Nov., 1981, p. 37-38.

33. Townsend, D.L., and Csathy, T.I., "Soil Type in Relation to Frost Action," Ontario Department of Highways, Ontario Joint Highway Research Program Report No. 15, Civil Engineering Report No. 29, 1963, 69 p.

34. U.S. Army Corps of Engineers, "Soil and Geology -- Pavement Design for Frost Conditions," Department of the Army Technical Manual TM 5-818-2, 1965.

35. Vlad, N., "The Determination of Frost Susceptibility for Soils Using a Direct Testing Method," Proceedings, Second International Symposium on Ground Freezing, Norwegian Institute of Technology, June 24-26, Trondheim, 1980, p. 807-814.

36. Zoller, J.H., "Frost Heave and the Rapid Frost Heave Test," Public Roads, vol. 37, no. 6, 1973, p. 221-220.

SUBJECT INDEX

Page numbers refer to first page of paper.

AUTHOR INDEX

Page numbers refer to first page of paper.